THE LEAN TOWER OF IT

The Concise How-To Guide to Implementing Lean Concepts to Achieve a World Class IT organization

Ray Jarrett

THE LEAN TOWER OF IT

The Concise How-To Guide to Implementing Lean Concepts to Achieve a World Class IT organization

Written by Ray Jarrett,
Overland Park, Kansas.

THE LEAN TOWER OF IT

Copyright: 2012 by Lean Strategies. All rights reserved. Printed in the United States of America. Except as permitted under the United States Copyright Act of 1976 no part of this publication may be reproduced or distributed in any form or by any means, or stored in a database or retrieval system without the prior written permission of the publisher.

ISBN: 978-1475241921

Title ID: 3858577

All trademarks are trademarks of their respective owners. Rather than put a trademark symbol after every occurrence of a trademarked name, we use names in an editorial fashion only, and to the benefit of tile trademark owner with no intention of infringement of the trademark.

Amazon.com books are available at special quantity discounts to use as premiums and sales promotions, or for use in corporate training programs. To contact a representative please order online.

CreateSpace books are available at special quantity discounts to use as premiums and sales promotions, or for use in corporate training programs. To contact a representative please e-mall rjarrett@newleanstrategy.com.

This publication is designed to provide accurate and authoritative information in regard to the subject matter covered. It is sold with the understanding that neither the author nor the publisher is engaged in rendering legal, accounting, or other professional service. If legal advice or other expert assistance is required, the services of a competent professional person should be sought.

- *From a Declaration of Principles jointly adopted by a Committee of the American Bar Association and a Committee of Publisher*

TERMS OF USE

This is a copyrighted work and Lean Strategies, LLC and its licensors reserve all rights in and to the work. Use of this work is subject to these terms. Except as permitted under the Copyright Act of 1976 and the right to store and retrieve one copy of the work, you may not decompile, disassemble, reverse engineer, reproduce, modify, create, derivative works based upon, transmit, distribute, disseminate, sell, publish or sublicense the work or any part of it without Lean Strategies prior consent. You may use the work for your own noncommercial and personal use: any other use of the work is strictly prohibited. Your right to use the work may be terminated if you fail to comply with these terms.

THE WORK IS PROVIDED "AS IS." Lean Strategies MAKE NO GUARANTEES OR WARRANTIES AS TO THE ACCURACY, ADEQUACY OR COMPLETENESS OF OR RESULTS TO BE OBTAINED FROM USING THE "VORK, INCLUDING ANY INFORMATION THAT CAN BE ACCESSED THROUGH THE WORK VIA HYPERLINK OR OTHER'WISE, AND EXPRESSLY DISCLAIM ANY WARRI\NTY, EXPRESS OR IMPLIED, INCLUDING BUT NOT LIMITED TO IMPLIED WARRANTIES OF MERCHANTABILITY OR FITNESS FOR A PARTICULAR PURPOSE.

Lean Strategies do not warrant or guarantee that the functions contained in the work will meet your requirements or that its operation will be uninterrupted or error free. Neither Lean Strategies shall be liable to you or anyone else for any inaccuracy, error or omission, regardless of cause, in the work or for any damages resulting therefrom. Lean Strategies has no responsibility for the content of any information accessed through the work. Under no circumstances shall Lean Strategies be liable for any indirect, incidental, special, punitive, consequential or similar damages that result from the use of or inability to use the work, even if any of them has been advised of the possibility of such damages. This limitation of liability shall apply to any claim or cause whatsoever whether such claim or cause arises in contract, tort or otherwise.

Table of Contents

Chapter 1: The Lean IT Blueprint	3
Common goals for IT and the Voice of the Customer	3
Digging Deeper	4
Chapter 2: World Class IT – A Foundation of Excellence	7
The Goal of IT	10
The Evolution of IT - Implications for Your Organization	12
And the Goal is…	15
Supporting Goals with the Right Metrics	16
Chapter 3: Understanding Lean and Inherent IT Pitfalls	19
Failing at Lean	22
So what is Lean?	26
The Pursuit of Perfection in a Lean Culture	29
Chapter 4: A New Model of Prefabricated Parts	31
The Emergence of Patterns	32
Work Categories and Lean Patterns	33
The Completed Blueprint	36
Headed in the right direction	38
Chapter 5: Pattern Examples and Deployments	41
R.P.M. Pattern - Pattern Ranking Score = 30 to 45.	41
R.P.M. Pattern Examples	44
R.P.M Review	53
P.I.C. Pattern - Pattern Ranking Score = 16 to 25.	54
P.I.C. Pattern Examples	58
P.I.C. Review	61
T.O.N.E. Pattern - Pattern Ranking Score = 0 to 15.	62
T.O.N.E. Pattern Examples	65
T.O.N.E. Review	67
Chapter 6: Case Studies and Tools	69
R.P.M. Pattern Case Study	70
P.I.C. Pattern Case Study	77
T.O.N.E. Pattern Case Study	90
Chapter 7: Putting IT all together…Are We Winning?	97
Our Journey in review	97
When is the Journey Complete?	101

Chapter 8: The Voice of the Customer…Revisited	103
Chapter 9: KATA – Implementing a Lasting Lean Culture	107
About the Author	113
Bibliography	119
Index	121

Introduction

What do you see?

Whenever I hear those words associated with a picture of some sort, my mind immediately conjures up images of old movies where an elderly, wise looking psychiatrist is showing some skeptical subject Rorschach ink blots. My favorite part is where the doctor says, "Umm-hmm, very interesting."

Some people look at this picture and see a sunrise. Others look at it and see a sunset.

What do you see?

While working at a leading Aerospace and Defense company, I had the privilege of selecting and hiring quality IT and Supply Chain professionals. As with most Fortune 500 companies, each new employee's first stop is orientation. Since we were primarily a manufacturing site, part of the orientation follow-on was a class in Lean Manufacturing. This class gave participants an overview of what they could expect once they started working in their respective areas. The class covered concepts like kaizen, heijunka, visual factory, andon, and 5S. The instructor urged each participant to take ownership for improving the metrics in their respective work areas.

At the end of these Lean classes, all of the new employees' managers and directors received invitations to a lunch meeting where each employee had 5 minutes to present a problem they were going to tackle using the tools they learned in the class. It was clear that some participants in the class saw something different than others. Most of the employees who worked in the production areas had clear examples of how the tools applied to their specific situations. The office workers seemed far less sure of themselves and how to apply the tools to their problems.

I decided to follow up on how things were going with one of the other production workers I'd met in the aforementioned meeting. I had often seen this particular employee working during my visits to the operations floor.

The production manager showed me pictures his employee took of an issue needing resolution. Afterwards, he explained in detail how changing the production process would result in a 5 minute per part reduction in assembly time. The new process would reduce the chance of scrap and subsequent rework by almost 100%. This left only variations in the raw material as potential show-stoppers.

When I followed up with my new employee, it was evident that she *failed to see* (through no fault of her own) how any of the tools could be utilized to make improvements without causing her to do MORE non-value added work. One person saw a sunrise while the other saw a sunset.

This caught my attention but the real "ah-ha" moment came when I tried to introduce Lean to the IT department. I gathered my management team together, announced my intentions, and sent them to a 3-day hands-on class. I asked them to give me 2 resources from each of their areas to serve on the team responsible for driving change in the organization. These would be the "Lean experts" who would undoubtedly prompt an infectious change and a Lean culture in our organization.

This turned out to be a huge disaster! The "team" showed up to each meeting unmotivated and unenthusiastic. They all viewed it as a huge waste of time. I selected the set of metrics I planned on reporting to executive management and they supplied me with the supporting data and the graphs. Cursory attempts at improvements were done without the use of the tools or any root cause analysis to determine that we had really fixed the problem. I could go on about everything we found wrong, but you get the idea. More on this story later…

After about 3 months of this, I disbanded the team and decided I would broaden my education in Lean, consult with some experts and determine where I'd made my missteps. After looking at Lean from many different sides and reading the combined works of many experts in our field, I began to notice some patterns. I also started hearing the same stories over and over again. "We don't have that kind of time," "We already are doing something like this" and "These concepts don't fit in a knowledge organization" were common statements made by many of the IT professionals I'd met and talked to. Our little corner of the world was not alone; many organizations were facing similar problems in their quest to implement Lean in a non-manufacturing setting. After accumulating and formulating a different way to look at and implement Lean IT, it seemed appropriate to write it down and share it.

I had a blast writing this book. I had a chance to visit with colleagues I'd worked with in the past and get opinions from seasoned, well-respected professionals. I had an opportunity to read all of the books and articles I was suppose to read and I made several trips down memory lane as I sifted through the thousands of spreadsheets, presentations and white papers I'd written over the years.

My intent was to use as many original artifacts as possible to show the journey of discovery towards arriving at these conclusions. You will see a number of pictures of dry erase boards, bulletin boards and meeting notes. If your goal is to create a culture focused on execution, use the concepts in this book. These approaches are not based on theory alone; they are battle-tested concepts that yield results.

You will get information on <u>how to</u> implement a Lean culture in your IT or knowledge work organization. This book introduces the concept of Pattern Matching – the idea that certain types of knowledge work fall into categories that lend themselves to certain types of Lean practices. What you will also get are visuals, practical examples, and case studies that have been recreated so that you will have a much easier time *seeing* the benefits of Lean IT.

And by the way… what you *saw* was a picture of a sunset!

The Lean Tower of IT

Chapter 1: The Lean IT Blueprint

Common goals for IT and the Voice of the Customer

The construction of the tower in Pisa began in the year 1178. In the first five years of construction, the building started to sink on one side due to the fact that the foundation was only three meters deep and seated in weak and unstable soil. The interesting fact is that over the  next 172 years, they continued to build on a foundation they knew would never be right. After many attempts at reinforcing the tower with weights and dirt removal, it was stabilized at its current tilt of 3.97 degrees. (Brittanica.com, 2009)

While the concept of Lean Manufacturing has been around a while, Lean implementations within IT are new and they end up a lot like the tower; built on a sinking and unstable foundation. Most Lean IT implementations are built on a few of the Lean concepts surrounding waste and the seven deadly sins; however, a true system of pull that represents the customer's order, facilitates flow, and reduces waste never materializes. Consequently, the culture is never created and the processes are difficult to maintain. Lean IT ends up being built on a shaky foundation that never rights itself.

The deployments I've seen and discussed with other technology leaders are one-off implementations of Lean Manufacturing and the seven deadly wastes sprinkled with a bit of Lean jargon. Simply put, a different set of blueprints and tools are needed to build a "Lean IT" house. The principles and concepts need not change, but the strategy may need a fresh perspective. Architects and builders would not dream of using the same tools, materials, blueprints and codes to build an office complex as they would use to build a housing subdivision. We do note; however, both endeavors require similar things such as plumbing, electrical and central air/heating.

Digging Deeper

After taking trips to other Lean Manufacturing sites and reading several books and articles, I came to the conclusion best summarized in Wikipedia's narrative on Lean IT,

> "*Although Lean principles are generally well established and have broad applicability, their extension from manufacturing to IT is only just emerging. Indeed, Lean IT poses significant challenges for practitioners while raising the promise of no less significant benefits. And whereas Lean IT initiatives can be limited in scope and deliver results quickly, implementing Lean IT is a continuing and long-term process that may take years before Lean principles become intrinsic to an organization's culture.*" (Wikipedia, Lean IT, 2012)

Most of us have discovered several disassociated tools, concepts, and disciplines but nothing representing a "game changer" having the potential to impact all IT disciplines and other knowledge work organizations. A good IT related example is Object Oriented Programming (OOP). When OOP emerged as a concept, it spawned a new wave of concepts related to technology culture such as objects, UML and Use Cases. Tools such as Strategy Pattern languages like VB.Net and C# as well as the Adapter Pattern language Java, forever transformed the face of application development. Lean IT has the same potential; however, unlike its manufacturing predecessor, it lacks a culture and a foundational set of goals which would normally give rise to tools, standards, blueprints and templates for IT – *as a whole*. We see significant Lean contributions in the arena of application development. Specific tools such as KANBAN (pioneered by David Anderson), the Agile Manifesto authored in 2001, and the Declaration of Interdependence (based on the Agile Manifesto) are examples of a small cross-section of Lean's impact on the total spectrum of IT disciplines. Lean IT is truly just emerging.

Chapter 2: World Class IT – A Foundation of Excellence

Many architects and builders eventually settle on a standard set of blueprints when constructing office parks and subdivisions to avoid starting from scratch each time they undertake a new endeavor. It seemed as though it would be impossible to find a standard,  foundational blueprint in such a diverse landscape of IT organizations. That is, until I had the pleasure of meeting Peter High. Peter is the President of Metis Strategy and author of the book, "World Class IT: Why Businesses Succeed When IT Triumphs." After talking and consulting with Peter, it is clear he is a thinker, a strategist and well-respected in technology circles. His book is a must-read for any serious IT executive, leader or strategist. His framework introduces a common set of principles each IT organization should embrace and cultivate to meet the goals set out by their respective businesses.

I highly recommend Peter's book - add it to you library. Peter's basic principles can be summed up as follows:
1) Recruit, train and retain World-Class employees,
2) Build and maintain a robust IT infrastructure,
3) Manage projects and portfolios effectively,
4) Ensure partnerships within the IT department and with the business, and
5) Develop a collaborative relationship with external partners. (High, 2009)

One of the many attractive features of this model is its inherent timelessness. This framework can be easily deployed to ensure any company's ultimate success. This model serves as the perfect baseline for each IT department to focus on prior to implementing the specifics of Lean.

It only makes sense to have a firm foundation of "behind the scenes" IT functions before attempting any endeavor built on these services. These are services users and customers expect to be in place in any IT organization. How can we teach Lean principles to IT project managers and business analysts if we do not have the right people in place? What if there are no solid prioritization strategies or governance agreements in place with the customers? It hardly seems appropriate to start a value stream map of a process if it is not supported by the right people, frameworks, portfolio management disciplines and collaborative relationships. These must first be in place before beginning an effort to execute and monitor any new Lean processes.

The World Class IT (WCIT) model is a solid foundation on which to base new Lean thinking. Each principle in this model is comprised of measurable sub-principles and tools the IT executive can quickly put into place. Organizations and IT entities must know the following before proceeding with a Lean Implementation:

1) What is the business expecting from IT? What is the overall business strategy of the corporation or group and how does IT fit in?
2) What do the customers (internal and external) need from IT? What is the voice of the customer?
3) How will you measure your success? What metrics correlate with the business and customer needs? How can you or anyone else tell if IT is winning or losing?

Many Lean implementations fail because IT organizations are trying to base a new culture on an unstable foundation of basic services. Why would a football team try to run a "Cover 2" defensive scheme when they have not mastered basic blocking and tackling? (I have officially thrown in my obligatory sports analogy.)

If these basic services are not in place, a Lean Implementation is sure to be met with internal resistance from the professionals who are attempting to bridge the gap between their "real jobs" and what is certain to be viewed as the "process flavor of the month." Lean's intent is to change the culture and become the new way of thinking and problem-solving.

The Goal of IT

What is the goal of IT? This is undoubtedly a big question, but other business units such as Finance, HR and Supply Chain have very clear, universal disciplines and goals in place. If we can frame a definition and a clear set of well-vetted IT goals, we can begin to understand effective ways to measure and consistently apply Lean principles. As with any seasoned Lean implementation in the manufacturing sector, IT leaders should be able to see measurable, quantifiable financial gains, cost savings and increased business capabilities.

In most manufacturing operations organizations, the goals and objectives easily flow down from the business – they start with defined production goals and end with measurable metrics such as parts per minute for the workers on the assembly line. If the business determines that there is a need to increase production by x% to fill a backlog of orders, there are levers in the form of goals and metrics the production leadership team can pull to eliminate waste, increase flow and reduce scrap. This reduction in waste is just one example of using Lean to achieve the desired goal.

But what set of goals and metrics does the *IT manager* monitor to increase production by x%? How do we determine if our systems are meeting the needs of the business? Moreover, how do business leaders make decisions to cut or increase yearly IT spending? What overall metric tracks the success or failure of the IT department and do those metrics tie directly to the production of goods and services or the profitability of the company? What baseline metric indicates the need for improvement? At any point, we should know when we are winning or losing the battle. If our production counterparts know the answers to these questions with regard to their operations, we should know them about ours as well.

Each different company's IT functions and goals are as widely varied as the Midwest's definition of "good barbeque." The focus of some IT organizations is with Keeping the Lights On (KTLO). Others are primarily seen as a group of individuals who write custom code to fill in technology gaps within the company's suite of applications. In other businesses, IT groups are primarily viewed as call center and PC support organizations for other knowledge work groups within the corporation.

Fortunately, in recent years an emerging set of standards have provided some standardization in our space. Standards like COBIT, ITIL, TOGAF and SOA can serve as a common reference point, basis of governance and convergence in our industry; however, it still leaves our question unanswered…Can there be one, universally defined goal for IT organizations?

If we can create or adopt a clear structure and standard set of goals for our IT departments, we can distill those goals into patterns and actionable, measurable goals that fit within the context of Lean principles.

The Evolution of IT - Implications for Your Organization

"And remember, no matter where you go, there you are." **Confucius**

My favorite visual representation is the widely recognized, "You are Here" map. I have often suggested that IT organizations display the map in a prominent place so they can chart their progress. Over the past 20-30 years, we have witnessed the evolution of IT by observing the following phases (Figure 2.1):

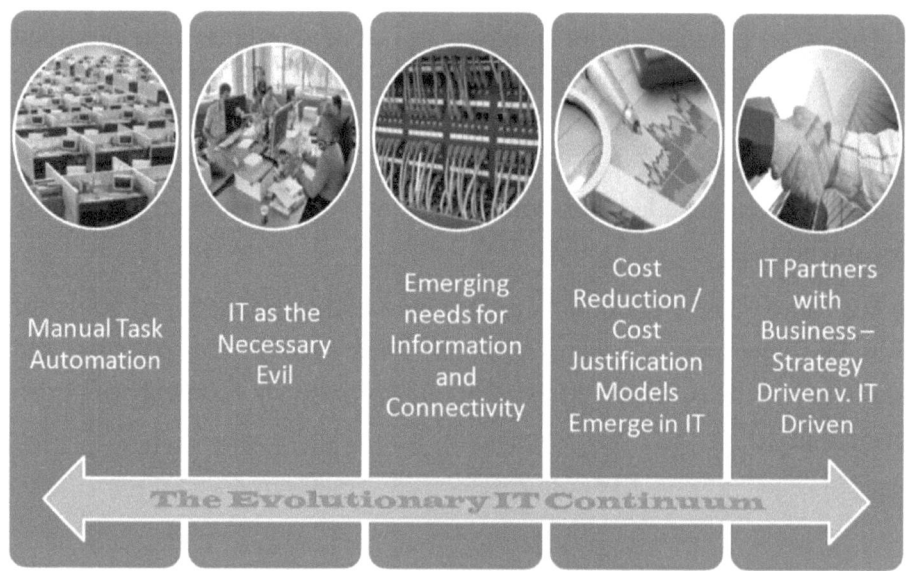

Figure 2.1 – Evolutionary IT Continuum

While many of my contemporaries posed good arguments for other categories or sub categories within IT's evolution, this text settles on the following eras since they resonate with the business community. Each phase was defined by the following characteristics:
1) Manual Task Automation – Businesses were focused on making their initial purchase(s). IT was called "DP" in the 70's and early 80's and applications were relegated to operational support levels of

automating administrative tasks. The IT budget was growing as fast as sales.
2) IT as a necessary Evil – Hardware, software and personnel were acquired *en masse*. Executives exerted little control over the growing, burgeoning, IT budget and formal project or cost justification was non-existent.
3) Need for Information (not just data) and Connectivity – COTS emerges as a viable alternative to the seemingly endless queue of development projects and the overhead associated with them. Data is decentralized and information is in constant demand. This forces the emergence of networking, security, data warehousing and ad-hoc request fulfillment. Architectures serve as templates for solution deployment.
4) Cost Justification and Cost Reduction – IT is now forced to cost justify projects and business units start to become more autonomous and force the de-centralization of certain IT functions.
5) IT leaders are forced to trim KTLO functions in their budgets and IT is driven more by strategy than technology.

An essential part of Lean is to determine where the organization is on the evolutionary scale. It is important to know if your organization is in one of the earlier stages of evolution. This knowledge will in all likelihood affect your approach to providing support and determining your metrics. Imagine the train wreck an organization is headed for when the business is focused on gaining a competitive advantage through the proper selection of software tools but the IT department is still in the early stages of getting their arms around basic support issues. Conversely, IT leaders cannot force the hand of the

business by engaging in the deadly waste of overproduction. If there is no vision or voice for services beyond KTLO, any blueprint or set of goals and metrics supporting anything beyond that will likely be met with resistance from senior leadership.

And the Goal is...

The goal of IT must be tied to the goals of the business, its support and ultimately its profitability. It is essential to have IT leaders who understand this and can bridge the gap between business and technology. Skilled dominos players have a saying, "all money is not good money." This means it is not necessarily a good strategy to make a play to score points each time the opportunity presents itself. By the same token, a skilled technologist does not deploy every strategy and technology presenting itself in the marketplace with little or no consideration for where the business is headed. The voice of the internal / external customer must always be heard, understood and *correctly* extrapolated.

Taking all of this into consideration, we can arrive at a goal to fit almost any IT organization, regardless of where it falls on the evolutionary continuum:

The goal of every IT organization should be to enhance the business' competitive position through the appropriate use of technology.

Supporting Goals with the Right Metrics

Most of our culture is based on winning and losing, or more accurately, on keeping score and measuring progress. In the last chapter, we ended our discussion by emphasizing the importance of metrics. I will refer back to the previous construction examples.

Before undertaking the expense and tedium of building a house, most people insist on a blueprint, a firm itemized cost estimate, a project plan and some assurance of quality verified by professional inspectors. Why do we expect these things? We want to know the score, especially when our hard-earned cash is the ultimate prize.

These are all ways of measuring progress. Without the right metrics in place, IT organizations have no way of knowing if they are meeting the expected goals of the business or if they are appropriately responding to the voice of the customer.

While it is vitally important we measure the right things, it is *paramount* we understand what the customer wants us to measure and what they want to ignore. Some IT organizations measure things amounting to little more than minimum entrance criteria. This is akin to tracking how often customers get hot food at a restaurant, or how many times a clean room is available to travelers checking into a hotel. We would think that an ice cream store had wasted its hard-earned advertising dollars for touting the fact that 99% of their customers were served cold ice cream last month. Similarly, which of our IT customers cares that the servers are up 99.999% of the time? IT customers are more inclined to want assurances that their PCs and applications are always working and that they have the right tools they need to solve problems. They care about how long they have to wait for solutions and projects to be implemented or if requests to the help desk are being correctly assessed and immediately addressed. CIOs care about cost containment and whether or not IT is helping the business to gain a competitive advantage. We must measure the right things if we are going to keep our seats at the table.

Chapter 3: Understanding Lean and Inherent IT Pitfalls

In this chapter, we will discuss the inherent issues associated with Lean implementations in IT and knowledge work organizations.

At this point, we have the makings of a good blueprint. We have established the need for a good IT foundation. The WCIT model provides us with a solid framework for every IT organization. We have also discussed the need for clearly understanding what the business requires from our IT organizations in order to meet shareholder expectations and financial goals. Meeting these goals will assist the business in gaining a competitive advantage in the marketplace.

Lastly, we covered the need surrounding our ability to turn goals into metrics that capture the voice of the customer. We must remember that some metrics are important, but they may fall in the category of minimum entrance criteria. Our internal customers expect us to keep the servers and PCs up at all times. They expect to have an infrastructure that supports their ability to accomplish their goals and they expect us to hire, retain and develop the talent we need to do these things.

The *right* set of metrics will reflect our success in the areas that matter to the internal / external customer and the business on an ongoing basis. Once we have the blueprint established and the foundation laid, we are ready to use Lean concepts to construct our "dream home."

There are several very well written books on the subject of Lean and I highly endorse them. One of the standard bearers on the subject is the book titled, "The Toyota Way – 14 Management Principles from the World's Greatest Manufacturer," written by Jeffrey K. Liker. Another must read is "Toyota KATA," by Mike Rother. I recently read "The Lean Startup" by Eric Ries – an excellent book on starting up businesses and organizations with Lean principles. I would be remiss in not offering up anything Don Reinertsen and David Anderson have authored. I highly recommend reading these books; they are an excellent "Lean starter kit" for your library. Before we delve into some of the key Lean concepts, it is vitally important to point out one of the major barriers to success with any Lean IT implementation.

In his book, "The Toyota Way," Liker stresses an important difference between using Lean tools versus creating a Lean Culture:

> *"After studying Toyota for almost 20 years and observing the struggles companies have had applying Lean manufacturing, what these Toyota teachers (called sensei) told me is finally sinking in. As this book attempts to show, the Toyota Way consists of far more than just a set of Lean tools like "just-in-time."* " (Liker, 2004)

To illustrate this point, all the reader needs to do is to briefly relive the all too common experience of calling Company XYZ's technical support hotline. Every now and then, we get the impression that the person on the other end of the phone is not really listening and probably does not really understand our problem. We sense the technician is reading a set of instructions from a card or script that hopefully will lead to a resolution. Contrast this with the expert who knows the intricacies of the application or piece of hardware we are calling about. This person knows when and how to apply this knowledge to each situation to create a customized answer to the problem. They know when to go "off-script." Similarly, Lean is not just a set of tools and instructions that will magically make all of our problems disappear.

There is a big difference between implementing Lean concepts and creating a Lean Culture. This difference explains why IT professionals have had such a difficult time with a concept that stands to yield such clear-cut results. Whenever the concept of Lean fails to catch on – the culprit is likely attributed to the fact that the organization failed to create the proper culture. Even in instances where there is a marked improvement in service or a particular product, oftentimes there is an equally marked degradation over time whenever culture is not created. In order for Lean to succeed, a culture of continuous improvement must be created.

As businesses and organizations changes, any previously implemented Lean process may need to be re-visited or changed in favor of something else that fits the new role or new customer demand. This type of trigger is not intuitive in an organization where culture is lacking. We must engage each worker in a way that unleashes their potential to be problem solvers who constantly scrutinize the flow of work in order to eliminate waste.

Lean implementation is not a new set of tools issued to employees with a new set of instructions for them to go out and "do Lean." Turning our employees into automatons is not the goal of Lean. I can best illustrate this point by telling the story about failing at my first attempt to implement a Lean culture.

Failing at Lean

Like most IT professionals, I am fascinated with learning new concepts, figuring out how things work, and fixing anything that is broken. Quite naturally, I was captivated with the whole Lean story and its fascinating new vocabulary - words like sensei, kaizen and... well, let's just say I was all ears. After my company bought me books and sent me to school, I was ready to add Lean thinking to a tool box already stuffed with methodologies such as Business Process Improvement (BPI), Management By Walking Around, Six Sigma, Juran jargon, Demming indoctrination, Ishikawa diagrams and a host of other things.

Armed with my latest certificate, I rushed back to my office and scheduled a meeting with the IT managers and supervisors. I briefly explained some of the Lean / TPS concepts and scheduled each of them in the next class. Additionally, I asked each leader to select a person or two from their staffs to participate on the new IT Lean Team. We sent the team to class as well so they could start working on my idea for an "Online Genba Board." I held an all-hands staff meeting and explained kaizen. We set a goal for completing a certain amount of kaizen ideas each month. I also told them that we all needed to 5S the office area. "I" was doing too much. Does any of this sound familiar? What happened next was what is typically called an "epic fail."

For starters, the kaizen ideas were what you might expect from people who were forced to do them. It would be kind to say that the employees did not put a lot of thought into most of them. I thought to myself, "People are definitely not taking this seriously." In an office environment, the idea of 5S is helpful to some degree, but I always had a sneaking suspicion that cleaning one's desk, filing the papers and marking off an area where the eraser, keyboard and mouse went were not what Taiichi Ohno had in mind. These were definitely not riveting tales for the knights of the Lean roundtable.

But the biggest failure of all was the "Lean Team." At each scheduled meeting, they gathered in our conference room and stared at me with that, "ok-tell-us-what-you-want-us-to-do" look. One person even mused that she hoped the meeting could end a little earlier because she had a lot of *important* work to get done. I forged ahead and reiterated how important Lean was to the plant and to our organization. With a determined look I stated that Lean was here to stay. Eyes rolled.

I confess I already had some improvements in mind I wanted our group to focus on. But deep down, I knew Lean was about empowered workers improving their own processes and workflows. What I ended up doing was trying to steer them into seeing what I thought was their set of issues.

Anyone who has worked with IT professionals knows they hate to be patronized and "handled." This was not going well. In the end, I was politely told that I should just let them know what I wanted so they could do it. I dutifully laid everything out and scheduled weekly recurring meetings to follow-up on our progress.

Each subsequent meeting was the same. They would gather statistics and update "my" website and listen to me say a few Japanese words. At the end of the meeting, I would ask for ideas on how we could improve. The problem was that there was no "we" so the team would just give me the one-thousand-yard stare and tell me they would try harder next time.

But the story gets worse. There were rumblings among some of the others in the department. Some wondered why *they* weren't picked to be on the Lean Team. They wondered what made the others so special. The rest of them were glad they weren't picked. On top of that, rumor had it that the Lean Team thought the whole thing was a waste of time. They thought it was yet another task they had to do in addition to their growing list of responsibilities.

At my next manager's meeting, one person suggested that it might be a better idea to just write the justification for additional headcount rather than trying to reduce the workload by doing this "Lean thing." After about three months of this, I realized I was the one doing all of the work. I disbanded the team, pulled back all the 5S worksheets and called it a wrap. My first Lean implementation had failed.

In the meantime, a friend and co-worker of mine who worked as an area manager in the factory started seeing positive changes in his organization after introducing Lean. I went to "go and see" what he was doing differently. When I got there, he immediately told me about a small victory his team had that day. Apparently, there was a problem on the line. Normally the workers would start the line at the beginning of the day and focus on keeping it running until their shift ended. Their main objective was to meet a target number of widgets for the day. After learning the importance of avoiding scrap, one of the workers noticed that they were starting to produce bad parts. She summoned the courage to stop the line. The team got their supervisor, fixed the problem and quickly restarted production – all on their own!

He then took me to an area where every hour on the hour the team could visually see an updated board showing the status of their "pitches" (Takt Time goals). He tilted his head and confessed that he did not know what behavior he was driving, but at least it was visual and the sensei liked it.

Lastly, we went back to his office where he showed me some of the kaizen cards he'd received from his team. One was a hand sketched diagram of how the line could be reconfigured to reduce each worker's wasted movements. Another person submitted a suggestion to change the kanban so that it would be easier for warehouse personnel to notice the replenishment signal. These were all really good ideas.

My co-worker told me he was working with our local sensei to get a Lean manufacturing culture up and going and wondered how my coaching sessions with him were progressing. I told him that I had no such sessions because I felt that I understood the material well enough to run with it - and we both said it at the same time…

…Apparently not.

So what is Lean?

My experience taught me what Lean was not. I had yet to learn that we had to change the *culture* of our organization. Culture dictates that everyone understands the voice of the customer, how to recognize the seven deadly wastes when they are occurring and how to use Lean tools to remediate issues. Our IT leaders had to learn that the workers were the processes experts and should be the ones making the improvements. I learned that Lean was a culture that needed to be cultivated.

Webster's Dictionary and Wikipedia both define *culture* as, "The set of shared attitudes, values, goals, and practices that characterizes an institution, organization, or group." I understood that to mean that we all had to see Lean the same way.

What we had to see can be summed up in three of Lean's major tenets:
1) The "prime directive" in any Lean enterprise undertaking is the drive to eliminate waste.
2) It is important to create a one-piece flow based on customer demand. This activity inherently exposes waste in a process.
3) Everyone must strive for continuously improvement. When people look for opportunities to improve the process, untapped creativity becomes a thing of the past. Lean is perpetuated because a culture is created. Culture endures long after the founding fathers have left the organization.

Taiichi Ohno summed up the Lean enterprise system at Toyota as follows,

> *"All we are doing is looking at the time line from the moment a customer gives us an order to the point when we collect the cash. And we are reducing that time line by removing the non-value added wastes."* (Ohno, 1988)

It is obvious that Ohno pursued and cultivated this culture. Talented individuals who embraced the ideas were given more opportunities and challenges. They were the new leaders in perpetuating the culture among their peers and subordinates.

My experience in working with IT and other knowledge work professionals on Lean implementations has been met with mixed levels of acceptance. One of the prevailing patterns of thought opposing the implementation of a Lean culture in knowledge work organizations is that Lean is not applicable to "our kind of work." In his book, "The Toyota Way," Liker makes a keen observation that speaks to this point.

> *"In courses I have taught on Lean manufacturing, a common question is "How does TPS apply to my business? We do not maker high-volume cars; we make low-volume, specialized products" or "We are a professional service organization, so TPS does not apply to us." This line of thinking tells me they are missing the point. Lean is not about imitating the tools used by Toyota in a particular manufacturing process. Lean is about developing principles that are right for your organization and diligently practicing them to achieve high performance that continues to add value to customers and society."*
> (Liker, 2004)

All organizations should follow a repetitive cycle of steps designed to assist with creating culture. A Lean culture *practices and repeats* the following model (Figure 3.1) for each pattern of work.

Practice relentless reflection (Hansei)	Define the issues based on input from the customer.	Document the existing value stream or process.	Eliminate waste in the value stream or process.	Establish Pull – Facilitate flow.	Monitor Results with the proper metrics.
• Ensure every voice is listened to • Brainstorm	• Always know the voice of the customer • Only monitor what they care about • Know real work vs. minimal entrance criteria	• A3s, Kaizen Cards and Kaizen Events • Value Stream Mapping (VSM) • 8D	• Find ways to skip steps that do not add value • Look for the 7 Deadly Wastes	• Know the demand and pace • Use Kanban and Heijunka	• Make everything visual • Check for expected results • Report Labor and Cost Saving

Figure 3.1 – Outcomes of a Lean Culture

The Pursuit of Perfection in a Lean Culture

When a Lean culture is established and cultivated, the people relentlessly pursue perfection in order to sustain a process. So what does perfection look like? Is it attainable? In my opinion, no technology literary effort is complete without an obligatory Star Trek reference. In the Voyager episode, "The Omega Directive," Seven of Nine has this exchange with the ship's first office, Commander Chakotay regarding her personal quest for perfection.

> *SEVEN: I have been a member of this crew for nine months. In all of that time, I have never made a personal request. I'm making one now. Allow me to proceed, please.*
> *CHAKOTAY: Why is this so important to you?*
> *SEVEN: Particle zero one zero. The Borg designation for what you call Omega. Every Drone is aware of its existence. We were instructed to assimilate it at all costs. It is perfection. The molecules exist in a flawless state. Infinite parts functioning as one.*
> *CHAKOTAY: Like the Borg.*
> *SEVEN: Precisely. I am no longer Borg, but I still need to understand that perfection. Without it, my existence will never be complete.* (Lobi, 1998)

Many management pundits believe it is too costly to attempt to attain perfection. Indeed, if we are striving for this goal by constantly adding functionality the customer did not request, it is *muda*. Nevertheless, striving for perfection is a practice we should seek to incorporate in the culture we want to create. This is best achieved when all parts in the organization see things the same way and are functioning as one – just like Seven's Omega molecules. A culture of relentlessly looking to improve and pursue a better way defines perfection.

Chapter 4: A New Model of Prefabricated Parts

Our group restarted our Lean journey and began by concentrating on creating culture. During a Lean training class full of IT and finance professionals, one participant made the following observation:

> "I have no inventory to speak of. I do not create any products for my customers that would require me to level load. The customer does want something from me, and that something is an answer to the problem they are calling about. I cannot detect defects (defect = help desk gets a call back to say that the fix did not work) in my product (product = answer to the problem) until after it is delivered. I am sure it's just me, but I don't intuitively see how I can apply these manufacturing principles to my kind of work."

I remember that the instructors tapped danced around the issue until they came up with an answer that kept the class moving. That was the comment that prompted me to take action. There had to be a better way to introduce Lean to IT and other knowledge work professionals.

The Emergence of Patterns

During my research on Lean manufacturing, I started to notice that different Lean approaches and tools worked best when paired with certain types of work.

The idea of prefabricated parts has been around since the early 1800's. Builders used this practice to accelerate home building and focus on other aspects of construction and design. The goal of putting standard templates together is to assist IT professionals in creating a Lean organization without starting from scratch. Our counterparts in the manufacturing sector have taken advantage of this concept for years. Internet search engines turn up many websites that specialize in selling a vast array of charts, tools and even floor tape (for setting up Kanbans) to assist with many of the common manufacturing concepts related to Lean. These shortcuts allow the sensei and coaches to place more focus on creating culture instead of creating tools.

As I continued my work in other non-production organizations, I started to benefit from the successes IT and other knowledge work organizations were experiencing. Given the widespread misunderstanding of Lean by knowledge workers, it became even more imperative to develop principles that were a good fit for their type of work.

A number of years ago, I attended a training seminar hosted by the META Group in Canada on the subject of Infrastructure Pattern Matching (Adaptive Infrastructures). I was impressed by the approach and decided that if it could work for security and infrastructure, we could also use the concept to define Lean patterns for IT disciplines. This allows a workgroup to adopt a pattern and immediately utilize templates and tools that facilitate developing a Lean culture for that particular type of work.

Work Categories and Lean Patterns

The next step involves choosing a pattern(s) for the kind of work you routinely perform. Nearly all IT disciplines can be grouped into 4 different types of work and 3 different patterns. (Figure 4.1)

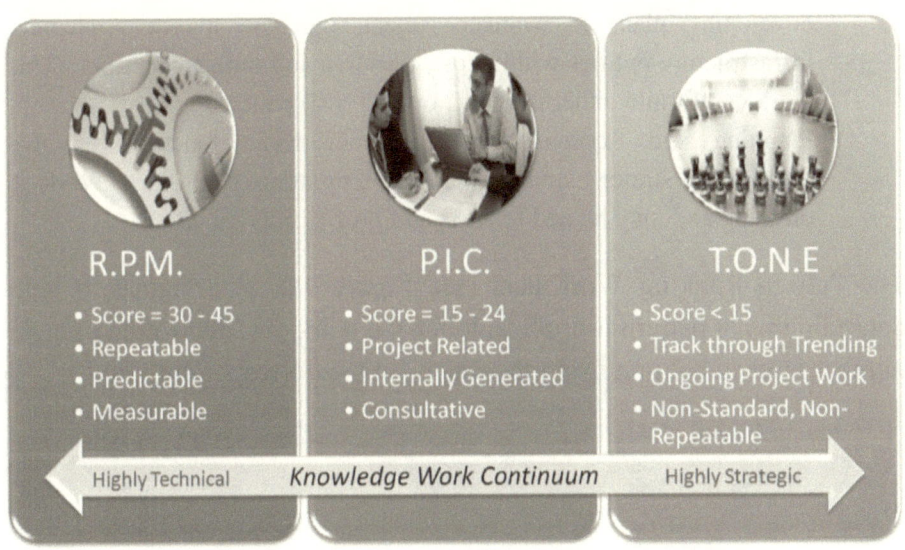

Figure 4.1 – Work Types and Pattern Matching Methodology

The Three Patterns are:
1) The R.P.M. Pattern – <u>R</u>epeatable Work, <u>P</u>redictable Flow, <u>M</u>easurable Output
2) The P.I.C. Pattern – <u>P</u>ortfolio Management, <u>I</u>nternally Generated Work, <u>C</u>onsultative
3) The T.O.N.E. Pattern – <u>T</u>rends, <u>O</u>ngoing, <u>N</u>on-Standard, <u>E</u>nterprise

Four Work Types:
1) Creative Work (producing widgets) – you create product for the customer by creating it from raw materials (hardware or prefabricated code stubs) or from your previous knowledge or training (from "scratch").

2) Maintaining and/or Monitoring Work – your basic function is to maintain or monitor the health and/or operating condition of any IT infrastructure entity. IT infrastructure entities are classified as hardware, software, networks, internal clouds, or any similar entity.
3) Implementation and Consultative Work – your basic functions involve developing plans to evaluate, select, purchase and/or implement infrastructure entities within cost, resource and time constraints. This work can fit into either the P.I.C or T.O.N.E. pattern.
4) Planning and Strategic Work – your basic functions involve developing strategy and conducting long term planning. Work is ongoing and may be tied to KPIs and profitability goals.

The Pattern Ranking Worksheet (see Figure 4.2) is designed to help determine what pattern is applicable for the work you do. Enter a single number in each column using the ranges in the corresponding row.

	Work Externally Initiated?	Is your work flow static?	Is work completed in Small Batch Sizes by >1 resource?	Can you set up a KANBAN of subcomponent?	Can you detect defects?	Can work be defined as a repeatable process?	Can this work be automated*	Work uses elements of Basic IT Principals**	Is your product a tangible deliverable?	Total
Mostly (rate 4 – 5)										
About 50% (rate 3)										
Seldom/ Never (rate 0 - 2)										

Figure 4.2 –Pattern Ranking Worksheet

The questions in the worksheet are:
- Work Externally Initiated?
- Is your work flow static?
- Is work completed in Small Batch Sizes by >1 resource? (more than one person)
- Can you set up a KANBAN of subcomponents? (PCs, parts, code stubs)
- Can you detect defects? (before delivery of product)
- Can work be defined as a repeatable process?

- Can this work be automated?
- Work uses elements of Basic IT Principals?
- Is your product a tangible deliverable?

Work with scores in the range of 30 – 45 would be classified as an R.P.M Pattern, 15 – 24 as a P.I.C. Pattern, and any work that scores less than 15 is classified as a T.O.N.E. Pattern.

With the work categorized into patterns, we can now focus on the types of tools that will help to facilitate a Lean culture in each work group.

The Completed Blueprint

The following diagram (see Figure 4.2) illustrates the completed Lean IT blueprint. IT organizations can build a Lean structure using the following model:

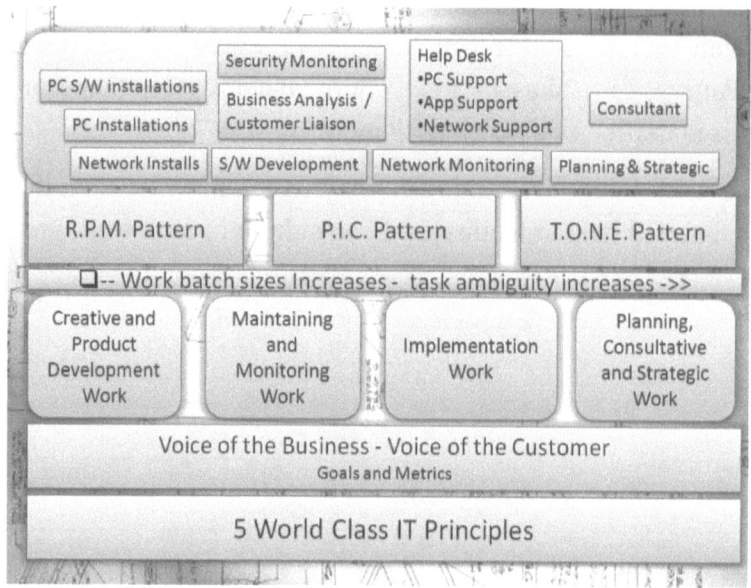

Figure 4.3 – Lean IT Blueprint

The blueprint can be broken down into the following sequential steps.

1. | 5 World Class IT Principles |

    The first step is to assess and monitor the organization's progress on the basics. The WCIT model helps the organization track and monitor progress with respect to:
    1) Recruiting, training and retaining World Class IT people
    2) Building and maintaining a robust IT infrastructure
    3) Managing projects and portfolios effectively
    4) Ensuring partnerships within the IT department and with the business

5) Developing a collaborative relationship with external partners

These basic building blocks are essential in ensuring that the organization has the right foundation on which to build a Lean infrastructure.

2. | Voice of the Business - Voice of the Customer Goals and Metrics |

The second step is to gather information on the voice of the customer and understand customer demand. This allows the IT organization to develop and track progress on the right metrics. It is important to understand where the business is headed and what it needs from IT to gain a competitive advantage. It is also key to know the customer demand. Do they expect zero trouble tickets in the queue? Is a 99% uptime goal for all applications reasonable? Do they want a 95% success rate on meeting install dates? Is a goal of zero defects realistic for any in-house developed application? Knowing exactly what the customer wants to track (they may already be tracking it) will ensure that you are focused on improving the right processes.

3. -- Work batch sizes Increases - task ambiguity increases -->>

| Creative and Product Development Work | Maintaining and Monitoring Work | Implementation Work | Planning, Consultative and Strategic Work |

Thirdly, it is important to categorize the distinguishing factors of your work. The work you do fits into a pattern characterized by a set of Lean tools that will facilitate the formation of a Lean culture in your organization.

4. | R.P.M. Pattern | P.I.C. Pattern | T.O.N.E. Pattern |

Lastly, select the appropriate pattern based on the results of the Pattern Matching Worksheet. Detailed explanations, examples and case studies are included in upcoming chapters.

Headed in the right direction

Armed with a new theory, we revisited the employee (and several others) who had questions about the validity of implementing Lean in our IT organization. We briefly touched on some of the concepts of the Pattern Matching Model by suggesting that he focus on understanding what his customers wanted and setting some goals for meeting their expectations. We asked him to try and make his efforts visual to all of the co-workers in his group. We told him that we would point out a few tools he could use to resolve issues afterwards.

Later in the week, most of the IT leadership team and the sensei were taking a "Genba Walk" to look at the visual signals and various continuous improvement projects. We saw a board consisting of a log of all of the trouble calls / tickets received and the number of successful calls vs. the calls needing "rework." I could tell he was not overly impressed with his efforts, but this was an admirable start. The data was worth exploring.

Essentially, the group was tracking First Pass Yield (FPY). That first spark in his eye let me know he had made a real connection with the information. I encouraged him for coming up with a measure that was important to the customer. By tracking FPY on a regular basis, he could see whether or not the group was doing the right things to delight the customer. I asked him to think about why certain trouble tickets required rework. Two days later this is what he sent me...

1) On some calls, we misunderstood the user's description of the problem.
2) In some cases, we asked the wrong questions and ended up resolving some other issue that may or may not have been related to the root of the problem.
3) In some cases, the user described some set of issues that were unrelated to the main problem.
4) We just didn't know the answer and had to do more research.

5) In some cases, the user was unable to recreate the problem and subsequently unable to explain it. This would require a call-back once the problem happened again.

This was exciting stuff indeed. I immediately introduced him to the "5 Whys" concept. Later, the employee partnered with some of his co-workers to help with the document. Afterwards, we did some value stream mapping and eliminated all of the waste in the process. This led to a recommendation to purchase a set of software tools that would ultimately improve FPY. This was great news - the employees in the group were really starting to catch on to Lean thinking. But what happened next was really exciting!

The next time I was in the area, I noticed that the board was no longer there. When I asked why, the employee told me that they were revising it to include some additional information. The 5 Whys and the root cause analysis got them to thinking about *why they were getting calls in the first place*. The team was now focused on grouping the calls into categories and triaging the issues so they could take measures to prevent problems from happening in the first place. This got them involved with other groups like desktop support and some of our external vendors. In the end, the group increased FPY to over 92%. They deployed software that allowed them to see the user's desktop. In addition, they developed standardized triage procedures and provided users with supplemental training. Additionally, they reduced the occurrence of trouble calls by over 35% by preventing the issues that generated problems in the first place. In the end, the IT manager was able to downsize the helpdesk group and redeploy the personnel to other areas in the IT organization. Some of redeployed individuals got promotions and other exciting career opportunities. The gaining groups and organizations now had individuals who were helping the Lean culture to spread.

It was great to see a Lean culture taking shape and the individuals making progress. More importantly, our new Lean approach made a difference in the quality of the services offered to the internal customers.

Chapter 5: Pattern Examples and Deployments

R.P.M. Pattern - Pattern Ranking Score = 30 to 45.

	Use Takt Time as a Metric	Use a Burn Down Chart	Set up the work components in Kanban stations	Document the process and look for continuous improvement (C.I.) opportunities	Set up, monitor and show quality metrics	Set up ANDON to stop the process when scrap is being generated	Make Work and Progress Visual
	Fixes Mura (斑 or ムラ) Unevenness	**Shows the state of the Work Progress**	**Kanban** (看板か) Also (かんばん)	Reinforces **Kaizen** (改善, C.I.) Elimination of **Muda** (無駄) also (ムダ) **Jidoka** (自働化) search for automation opportunities	Error-proofing **Poka-yoke** (ポカヨケ)	Elimination of **Muda** (無駄) also (ムダ) **Andon** (行灯) Signboard	**Genba** (現場) place where the real work is done

Rate 1-5 in each category; one score per column

	Work Externally Initiated?	Is your work flow static?	Is work completed in Small Batch Sizes by >1 resource?	Can you set up a KANBAN of subcomponents?	Can you detect defects?	Can work be defined as a repeatable process?	Can this work be automated?	Work uses elements of Basic IT Principals?	Is your product a tangible deliverable?	Total
Most of the Time (4 - 5)	4			4	5			4	5	22
About 50% (3)		3				3				6
Seldom / Never (0-2)			2				0			2
										30

Figure 5.1 - A sample Pattern Ranking Worksheet assessment from an in-house development group.

Most departments engaging in this type of work have a useable product that must be delivered to a customer in certain quantities on a specific date. The work in this pattern is defined by the need for specific, deliverable quantities, a certain pitch that must be maintained and a "kit" of raw materials (hardware or software) used to create the end product.

Take, for example, the department deploying 20 units per week as a part of a PC refresh project. Each PC must be loaded with a baseline image and augmented with specific software, drivers and peripheral devices germane to its intended user. In this case, there are specific sets of hardware and software components that go into producing a finished product. These components can be staged in a set of kanbans or they can be kitted separately.

A software development group that develops and deploys application software is another example of work in this pattern. Developers can build the applications with a number of pre-coded stubs and focus their attention on the specific logic and programmatic flow that make the applications unique. The due date gives the group a way to set their takt time and construct a burndown chart. In both examples, there is a definite, measurable way to break the work down into repeatable steps.

In each case, the use of "prefabricated parts" is highly encouraged. There should be quality checks and inspections as the product flows through the *construction process*. Completed parts and/or code should be moved to the product's final staging area. Progress can be tracked and measured in a burn down chart. Here are some other tools that work well within this pattern.

- Use Takt Time as a Metric. Takt time can be first determined with the formula: $T = T_a \div T_d$,
  Where:
- T = Takt time, e.g. [minutes of work / unit produced]
  T_a = Net time available to work, e.g. [minutes of work / day]

T_d = Time demand (customer demand), e.g. [units required / day]

Net available time is the amount of time available for work to be done. This excludes break times and any expected stoppage time (for example scheduled maintenance, team briefings, etc.).

> ➢ <u>Lean correlation:</u> Fixes Mura (斑 or ムラ) Unevenness

- Use a Burn-Down Chart – works in conjunction with Takt Time and visually represents where and what state the work is in.
- If there are several components that go into the finished product and it is possible for more than one person to work on the finished product, set up the work in stations where it can be used JIT.

  > ➢ <u>Lean correlation:</u> Kanban (看板, also かんばん) Sign, Index Card)

- Document the process and look for continuous improvement (C.I.) opportunities.

  > ➢ <u>Lean correlation:</u> Reinforces Kaizen (改善, C.I.), elimination of Muda (無駄, also ムダ), and Jidoka (自働化 search for automation opportunities)

- Set up, monitor and show quality metrics, i.e. % Scrap.

  > ➢ <u>Lean correlation:</u> Error-proofing Poka-yoke (ポカヨケ)

- Set up ANDON to stop the process when scrap is being generated.
- Create a Visual Board and set up daily or weekly visits to review.

  > ➢ <u>Lean correlation:</u> Genba (現場) - place where the real work is done.

R.P.M. Pattern Examples

The Hardware Group

The hardware and PC support department got feedback on their services in two different ways. The first was through an annual company survey given to all employees. Secondly, they received real-time information via IT surveys sent to a statistically correct sampling of users. This feedback showed that the department's (hardware and software) two biggest issues were:

1) Failing to correctly facilitate the new employee's on-boarding experience with the company. The goal was to have a phone, computer and cell phone (if applicable) available for them on their first day. Additionally, we rarely met requested delivery dates when an employee moved to a new department or added some feature / change to their configuration. If we were close to meeting the date, we certainly did not get the request right the first time, i.e., not all files moved over, wrong software installed, etc.

2) Failing to deliver architected software solutions in a timely manner. This was a very serious problem since ultimately it impacted our application infrastructure. Users were getting frustrated and attempting to create their own solutions with Access DB and Excel Band-Aids, with back door database queries that ultimately slowed down the online application, and by hiring third party contractors to come in and work directly for their department. (For the Software Group)

After meeting with our customers, the PC support team came up with the following observations, goals and metrics:

1) Our customers want all installs, moves, additions and changes to happen on the requested day with flawless execution. The only exception to this request was if IT received less than adequate notice. The customer defined "adequate notice" as 1 week for changes and 2 weeks for new requests.

2) The customer wanted to know if their request was received and being processed.
3) The customer wanted a way to see the status of the request via the intranet (online).

The department came up with specific metrics and the AS-IS condition. They chose "Percent of on-time installations" and "Percent of Flawless Victories" as metrics that mattered to the customer. They also drafted an A3 to initiate the project with a goal of achieving 90% for each metric (based on the AS-IS condition). A cross section of our users agreed that these goals were specific, measurable, attainable and realistic.

The A3 revealed other flaws in the process. There was no single, centralized location where customers could make all of their requests. There was no way to check the status of each request. There was no schedule, rhyme or reason as to when and how often patches and security updates should be applied. Several other flaws appeared, but the group reasoned that these fell in the range of "basic services" and that the customer would expect that these items were being done behind the scenes. These capabilities were in the foundational "blocking and tackling" areas of WCIT.

Following the Lean culture model in Chapter 3 (see Figure 3.1), the group used a series of tools to truly uncover the processes and how they worked. After an early attempt to map one of the processes, the group was not convinced that the actual, documented steps were being followed. The group decided to have an outside person, "go and see" by following the work process and capturing it through the eye of a video camera. What they saw was shocking! They counted over 30 undocumented steps in the process and nine different process branches off the main flow. There was so much waste that it was almost a guarantee that it was possible to exceed their initial 90-90 goals. One positive outcome was that they were able to "stopwatch" each step in the process. This additional task related information helped the group to uncover why there were such huge swings in task completion times between the different workers.

After eliminating waste (all non-value added activity) from the process, the team took on the task of leveling the work flow and establishing pull. They realized that part of their basic services to the customer were missing; specifically, the ability to track all relevant information about a customer's request. To fix this crack in the WCIT foundation, the team deployed a simple tool to allow any user to open a trouble ticket. The group deployed an aggressive mass marketing strategy and advertised a new service portal. Here, each customer could be assured that their requests were being reviewed and tracked. This eliminated any instances of dropped requests that came from phone calls, emails and voice mail messages. At the same time, they also agreed to a minimum lead time of 1 week for any equipment relocation or configuration change. Any new installation required a lead time of 2 weeks. This made it possible for the group to know the exact customer demand for the month.

This small change also allowed the group to construct a visual board and level load the work based on location and date. Instead of trying to manage a day where there might be 10 scheduled changes across 4 different locations, they leveled the work across an entire week in order to exceed customer expectations. In the early going while the team was acclimating to the new process, if an install was going to be a day or two late, they proactively called ahead and scheduled a new time with the customer.

This was a vast improvement from the customer's perspective since they no longer had to initiate a call to the help desk to determine the status of their request(s). This part of the Lean journey required the team to make a number of iterations to the new Heijunka Board in order to get things just right. Now, the technicians gather at the board each morning over coffee and can immediately see what needs to be done on any particular day. A card representing each individual customer request moved to the completed side of the board each time IT finished the work. The group deployed and maintained a real-time, burn-down chart to add even more visibility to the status of their works queues and metrics. It was easy for any passerby to tell if the group was winning or losing!

This addressed the production and waste problem, but the group still had persisting quality issues. At this point, the sensei introduced the concept of standardized work and error-proofing. Now that the processes were optimized, the group looked for new ways to test configurations and ensure that hardware parts worked perfectly (pre-testing). When moving a PC or installing new software, each technician diligently followed a pre-determined set of steps to complete the tasks. Not surprisingly, the discussions related to this improvement actually brought out some "best practices" for the technicians to follow. It seems that previously, each technician used their own knowledge and set of scripted notes to get the job done. This reduced the overall mean time to repair for each request.

The group constructed a quality board that tracked each ticket and showed instances of rework. These iterations took time to solidify but yielded notable gains in quality. Frequently, the group would shoot for a "Flawless Victory" week or month. These types of wins were very rewarding for the team.

These improvements led to an expanded scope and view of what was possible. The team worked with the managers and vendors to establish a better flow of hardware materials. They also partnered with HR to generate automated tickets and triggers so that when a new person was hired, all of their IT and electronic needs were captured. This way everything was set up and available for use on their first day. The group also theorized that they could eliminate calls by establishing an IT User Training Module to include in the new employee orientation class. All of these improvements came from the workers; management was no longer solely responsible for coming up with all of the ideas to improve service. The birth of a new Lean culture gave rise to drastically improved morale and customer service.

The Software Group

The Lean journey for the software development team was fraught with a few more lions, tigers and bears. At this particular worksite, the sensei created a very good training course on Lean, but it was entirely based on manufacturing examples and deployment strategies. In hindsight, it would have been a good idea to meet with the software developers to level-set their expectations prior to sending them to the course.

In our first sparsely attended meeting, the developers unanimously voted on the abolishment of Lean from the department. After attending the class, they were convinced that there was no way to apply the principles of Lean to their work. By this time, the book titled "The Toyota Way" started to appear on many desktops in the organization. I started quoting chapter and verse as to why this common misunderstanding found its way into many knowledge work organizations.

Fortunately, the lead software developer in the group refused to dismiss Lean concepts at face value and started to do some research on his own.

The lead had already compiled a list of customer concerns related to the software development (SD) group. The word "chaos" barely described the tenuous state of the SD union. There were over 200 requests in the projects queue. Some projects were over 2 years old. The group completed a handful of moderate-sized projects the previous year – not nearly enough to keep pace with the demand. Customers were downright irate and most of the organizations were either hiring consultants to do their software development bidding or relying on some internal resources to develop add-ons. We estimated the proliferation of non-sanctioned database tools had reached over 4,000 in a community of 200 users. We did not even bother to count Excel spreadsheets. We failed to meet delivery dates and programmers could not move on to other projects because of quality problems associated with previously completed projects. This was indeed a sad state of affairs.

The lead demonstrated that he had a good grip on the basics of Lean and introduced the concept of Agile Programming at our next manager's meeting. After bringing everyone up to speed on the concepts, the group collectively decided to start a pilot project.

Software development has long been viewed as an art form that should not be fettered with constraints that could potentially limit creativity. In the past, delays in software delivery dates were so nebulous; IT managers were typically allowed a carte blanche pass on meeting deadlines. Out of all of the IT disciplines, software development has the greatest influx of methodologies and tools that lend themselves to Lean adoption. With the advent of Agile and Scrum and the influence of noted Lean experts such as Reinertsen, David Anderson, James Sutton and Eric Reis, software developers are able to directly apply the Lean principles of pull, flow, Kanban, jidoka, and poka-yoke to their application development processes.

Two weeks and two books later, the developers invited me to a meeting. I was informed in the email invitation that I could not provide too much input since I was considered a "chicken" and not a "pig." I decided that I needed a little education on Agile if I was going to effectively play the role of a chicken!

There are a plethora of classes, books, seminars and how-to's on Agile. Suffice it to say, everyone was very impressed with the initial work of the SD organization. Two of the biggest and most important changes it brought to the department were smaller batch sizes and better quality control. With Agile's ability to deliver key functionality up front, the team tested and delivered smaller chunks of code at a regular, predictable pace. The customer heard from IT every week as opposed to once during project scope, and 6 – 9 months later when we were ready to conduct testing. We were sharper on our estimates and customer satisfaction improved drastically – the customer always knew where they stood and they had useable functionality, in some cases, three weeks after the initial charter and priorities were set. Agile certainly pulled its own weight in the Lean world.

There were; however, still some management rivers to cross to make Agile just as successful as the efforts in the hardware group.

IT already had executive management's support to start a Lean journey. The next thing the team needed was a business analyst, some software (for the help desk), and a little web development. We also decided to "over-communicate" by deploying a tactical advertising campaign directed at our customers. The campaign's main focus was to explain how we were changing our services in order to better serve their needs. The IT leadership team attended countless meetings with key stakeholders in order to inform them about our shared vision and goals. We tracked to a timeline that outlined how and when changes would take place. We asked users to re-validate and re-submit older requests to ensure that we had a valid queue of work. Over the course of the next year, we made significant, sustainable inroads towards a totally Lean culture. Management gave us their blessing to pursue this flurry of activity. Without their support, Lean would have definitely ended up as the flavor of the month.

Part of that support came in part through our ability to show proper due diligence. We outlined an aggressive path for ourselves during the first year of putting a new leadership structure in place. The department promised to deliver on 5 new initiatives that strengthened our base. Although Peter High's book on World Class IT was not on the bookshelves, each of these initiatives centered on perfecting the basics. We also proved that we were looking at the right metrics – we had over 350 responses to our customer survey; a more than adequate sampling from a population of about 500 support and production personnel. Cleaning up the queue, instituting a new help desk and reducing hardware ticket requests showed that we were serious in our efforts to create a sustainable Lean culture. We also consulted with marketing professionals on how best to transition our customer's thinking to the new paradigm. We wanted to change their perception of viewing us as a necessary evil to one of an engaged, progressive business partner. The IT leadership team seized upon every opportunity to speak at all-hands gatherings and staff meetings. We ran commercials on the scrolling monitors and posted information on the company intranet. In short, we intentionally painted ourselves into the Lean corner, got some quick wins, and built up some much needed momentum.

R.P.M Review

There are other IT disciplines that fit the R.P.M pattern and the same set of tools can be applied to develop a Lean culture in these departments. To review, the R.P.M. pattern lends itself to:

1) Definable process steps and value stream mapping.
2) Detectable waste elimination of non-value added steps in the value stream.
3) Easy adoption of standardized work lists and best practices.
4) Easily defined goals that are attuned to the customer's expectation.
5) The use of many different methodologies to shrink batch sizes.
6) The use of Kanbans and Heijunka Boards to establish pull and level-flow.
7) The use of burn down charts to track progress to a delivery date.
8) Easy Implementation of "Win /Loss" visual pictograms.
9) Metrics facilitate ease of reporting and tracking time and labor savings.
10) Quickly bringing new workers up-to-speed and taking advantage of their ingenuity to solve new issues.

Always remember to revisit any process when the business or customer's needs change. The key is not to stick with a Lean process through thick and thin but to continuously evaluate the processes - especially if conditions change.

P.I.C. Pattern - Pattern Ranking Score = 16 to 25.

- Make Work and Progress Visual
- Monitor Progress Using Gates
- Establish Consistent Gate Reviews both Internally and Externally
- Publish Governance and agree on a standard set of tools, templates and tactics
- Implement a quantitative risk monitoring tool and use it as an ANDON
- Establish a regular, set time to provide updates and visit the Gemba board

- Use % Complete
- Genba (現場) place where the real work is done
- Shows the state of the Work Progress
- Aizuchi (相槌) also (あいづち) Synchronizing your movements
- Nemawashi (根回し) Laying the groundwork; building consensus
- Elimination of Muda (無駄) also (ムダ)
- Andon (行灯) Signboard
- Genba (現場) place where the real work is done

Rate 1-5 in each category; one score per column

	Work Externally Initiated?	Is your work flow static?	Is work completed in Small Batch Sizes by >1 resource?	Can you set up a KANBAN of subcomponents?	Can you detect defects?	Can work be defined as a repeatable process?	Can this work be automated*	Work uses elements of Basic IT Principals?	Is your product a tangible deliverable?	Total
Most of the Time (4 - 5)					4	4		5	5	18
About 50% (3)	3	3		3						9
Seldom / Never (0-2)			1				0			1

28

Figure 5.2 - A sample Pattern Ranking Worksheet assessment from a business analyst group.

The score for the business analyst (BA) group and the PC support group are almost identical. If the BA projects are all similar, they could manage their Lean journey using the R.P.M. or the P.I.C. pattern. For example, a project management group that sets up franchise offices and stores would be in a position to use either pattern to foster its Lean journey. In this example, the analysis, cost, length of project and the project plan would be very similar if not identical.

The P.I.C pattern comes into focus as the work becomes more diverse and complex. The Pattern Matching Worksheet will always indicate a lower score for work that is highly variable and ad hoc in nature. Project management groups, IT Business Analysts, and Value Stream Managers would likely use this pattern. The P.I.C. pattern also works best for other knowledge work groups providing assistance to these groups. Any group engaged in consultative work, IT business relationship management, or tasked with finding synergies across an enterprise would benefit from the P.I.C. pattern.

Creating a Lean culture for the P.I.C pattern works best when the following tools are deployed:

- % Complete, R/Y/G visual triggers and boards.

- Monitor Progress Using Gates. Gates enable groups to have smaller batch sizes of work.

- Establish Consistent Gate Reviews both Internally and Externally. Gates can be used as quality checkpoints.

    ➢ <u>Lean correlation:</u> Aizuchi (相槌) also (あいづち) - Synchronizing your movements

- Work closely with key stakeholders to establish governance and agree on a standard set of tools, templates and tactics - This can be used as standardized work for P.I.C. pattern groups.

    ➢ <u>Lean correlation:</u> Nemawashi (根回し) laying the groundwork, building consensus.

> Lean correlation: Reinforces the elimination of Muda (無駄, also ムダ).

- Implement a quantitative risk monitoring tool and use it as an ANDON cord. There are many risk calculators available; one familiar tool is the FMEA (Failure Mode Effects Analysis) spreadsheet. Use one of these tools to determine the risks of each gate and the overall project elements such as On Time Deliver and Cost. Run the calculations often and track them visually. Your goals should be visual and bordered within reasonable upper and lower control limits.

    > Lean correlation: Andon (行灯) (Signboard).

- Establish a regular, set time to provide updates and visit the Genba board (will not be daily).

    > Lean correlation: Genba (現場) - place where the real work is done.

- For complex, highly specific, non-repeating engineering or new product development projects, I recommend a review of TRIZ - the Theory of Inventive Problem Solving (TRIZ is an acronym for the Russian word). This methodology can be used to identify differentiations in the types of work that IT professionals tackle. The basic premise behind TRIZ is that most problems follow certain patterns. If you are working in a business analysis or project initiation group where the work is highly variable, consider the advantages of a standardized approach. This will reduce the variability in the work process and improve work estimates. Although all 40 TRIZ principles are not applicable in the IT field, some will immediately lend themselves to functions / work within the P.I.C. pattern. Engineers who participate in some framework within the P.I.C.

pattern will certainly find that more of the principles are applicable, especially those that have more physical attributes associated with them, i.e. Phase Transition, Thermal Expansion and Strong Oxidizers.

It is important to note *that the propensity for waste increases proportionally as the batch size and amount of ambiguity in the process increases.* In Dr K. J. Youngman's web article "Why Do We Batch?" the author makes the following observation:

> *"Batching issues have a profound influence on the characteristics of any process and substantial gains can be made by properly understanding the dynamics involved. Although we often don't think about it, we can batch in either quantity of material or quantity of time. They seem interchangeable but most often one is treated as the variable and the other invariable. We batch once a week; means time is invariable and material is variable. We batch a full load; means material is invariable and time is variable."* (Youngman, 2009)

In the P.I.C. pattern, the objective and focus is on reducing the variability in time. Because of the larger batch sizes, the biggest areas of waste tend to be:

1) Waiting - waiting to hear answers back from the customer or vice versa,
2) Overproduction - the propensity for developers, engineers and project coordinators to add other tasks and features not required, and
3) Rework - resulting from unclear requirements and quality issues.

P.I.C. Pattern Examples

One thing the SD organization identified early on was the need to have a separate IT group working as a customer liaison. We hired a very talented business analyst (BA) after defining a new IT structure.

I mentioned the state of affairs with our backlog of development projects. Our new BA walked into a situation where there had never been any consistent standardization in the requirements gathering process. Her initial task was to distill the information already gathered from the survey and the development group with regard to the quality and timeliness of the solutions. The list she presented to management should look familiar to most IT managers:

1) We never get our software when it is promised.
2) We don't feel like the developers really understand what we do.
3) We have a language barrier – they don't have a clue what we are talking about nor do we know what they are talking about.
4) I know "that" is what we asked for, but why didn't you recommend something better?
5) We get software with features that we did not ask for and / or the features and functionality are not what we wanted.
6) I have no idea when my project will start, what priority it has, or how long it will take to complete.

The first concept the sensei introduced to the BA department was the concept of Genba; going to the place where value is created. The most painful reality of our organization was that we did not understand our customers and their goals. Many of our software developers had never been out to the manufacturing area to see our products being made. We immediately instituted tours through the main production areas. We never wanted to repeat the practice of starting a new development effort without seeing the need for it.

Our BA had a real advantage in that there was no process to change; she had a blank sheet of paper to fill with new Lean processes. Following

the Lean Culture Model, we established new processes to engage the customers, establish gate reviews, gather requirements, prioritize features, coordinate testing and deliver software. While all of this sounds unremarkable, the huge difference was in our ability to:
- Track our progress in relation to established metrics. (Are we winning or losing?)
- Institute and ensure quality standards for code functionality and delivery.
- Reduce the huge amount of waste in the process.

In order to standardize the process, we created standardized worksheets for ourselves and the customer by instituting governance and Service Level Agreements. We also took advantage of the smaller batch sizes Agile provides and established a calendar of standard meetings to:
1) Review Gates
2) Review work completed in an iteration
3) Test code
4) Train users, and
5) Release code into production

We used simple stoplight visual signals to show the state of projects. For some of the longer projects with financial implications for the internal customer, it was important to know the risks associated with potential delays, added features, etc. We used a visual risk tracking & probability system with upper and lower control limits to let the customers know the project was in control. IT pulled the andon cord and called a meeting with the customer representative whenever any time-sensitive gate or iteration was "yellow and trending to red." It was instantly visible to everyone whenever any particular task's risk or probability of failure started to meander past the upper control limits for more than a week.

This was radically different from the way we operated in the past and required a great deal of effort to level-set the customer's expectations.

It was necessary to successfully demonstrate how more frequent, smaller meetings and quality checks would result in a better quality product. Our up-front governance with the project sponsor stipulated the need for the user to commit to regular meetings in accordance with a pre-defined project schedule.

Lastly, our BA kept track of our new performance metrics and showed a marked improvement in quality (trouble tickets post implementation), project length, and rework (incorrect, missing or additional features). This further substantiated the positive contribution of Lean in our organization and helped to establish our reputation as viable business partners.

P.I.C. Review

The P.I.C pattern is specifically for processes related to project work, internally generated work and/or consultative work. These work patterns usually have large periods of time between gates. Lean strategies help to decompose these tasks into smaller batch sizes, eliminate non-value added steps and wasted time. The P.I.C pattern:
1) Focuses on eliminating the variability in time estimates. Reduces the occurrence of long task completion timeframes.
2) Reduces batching.
3) Helps to eliminate the three biggest problems in this work pattern:
    a. Waiting on responses from customers or vice versa.
    b. Overproducing by adding features or producing work not requested by the customer.
    c. Rework resulting from unclear requirements and quality issues.
4) Eliminates the unwieldy project plan as a visual. Deploys the use of gate reviews and metrics / measurements that delve into specific tasks that put the project at risk.
5) Makes use of quantitative risk analysis as an excavation tool to drive process improvement.
6) Implements a way to instantiate an ANDON cord for work patterns where this would have previously been impossible or impractical.
7) Drives the use of governance to lay the groundwork and build consensus with the customer. (Nemawashi) Governance is the standardized work of the P.I.C. work pattern.

T.O.N.E. Pattern - Pattern Ranking Score = 0 to 15.

- Focus on Cost as a Metric and track trends.
- Develop and Focus on Customer Satisfaction Metrics and track trends
- Develop a continuous feedback loop on Quality and track trends.
- Focus on %Complete (25-50-75-100)
- Ensure Enterprise KPI's are linked to Metrics
- RPM or PIC performance may be roll-up to your Gemba Board
- Use Kaizen Events and Lean Tools to triage difficult issues / problems and set up a P.I.C or R.P.M pattern until resolved.

Rate 1-5 in each category; one score per column

	Work Externally Initiated?	Is your work flow static?	Is work completed in Small Batch Sizes by >1 resource?	Can you set up a KANBAN of subcomponents?	Can you detect defects?	Can work be defined as a repeatable process?	Can this work be automated*	Work uses elements of Basic IT Principals?	Is your product a tangible deliverable?	Total
Most of the Time (4 - 5)										0
About 50% (3)					3	3				6
Seldom / Never (0-2)	2	1	0	0			0	2	2	7

13

Figure 5.3 - A sample Pattern Ranking Worksheet assessment from an Executive IT leadership team.

The power in this pattern lies in its ability to allow tasks with very large batch cycles, non-repetitive processes and strategic implications to become less susceptible to failure. The majority of work in this pattern fails if too much time elapses without noticing defects in the process responsible for circumventing success. In the manufacturing setting, rework is undesirable, but possible in order to avoid a total loss of investment. With T.O.N.E. work patterns, rework is highly unlikely or impractical; leaders rarely get a shot at a "do-over" with these goals. It is paramount that leaders quickly recognize problems and make timely pivots and course corrections. Smaller pivots and course corrections allow leaders the ability to ensure success by detecting defects before it is too late to fix them. Any assignment that monitors ongoing trends, has no definitive end date, is non-iterative or repetitive, and is tied to enterprise or cash/sales/profit goals are examples of work fitting this pattern.

An example of non-repetitive work would be the Supply Chain Director who is responsible for erecting a Warehouse on company property or dismantling the current warehouse and lowering inventory levels in favor of KANBAN / JIT shipments from suppliers. This work is best served by utilizing the following tools to monitor waste and waiting:

- Focus on Cost as a metric and track trend changes in the metric.
- Develop and Focus on Customer Satisfaction Metrics and track trends.
- Develop continuous feedback loops on Quality and track trends.
- Focus on %Complete (25-50-75-100).
- Ensure Enterprise KPI's are linked to Metrics.
- A subordinate's RPM or PIC performance metrics may aggregate up to visual metrics or a Genba Board at this level.
  - <u>Lean correlation:</u> Genba (現場) - place where the real work is done.
- Use Kaizen Events and Lean Tools to triage difficult issues / problems representing a subset of a larger initiative. This may show up as a "blip" on the radar during iterative trend monitoring. Monitor the work as a P.I.C or R.P.M pattern until resolved.
  - <u>Lean correlation:</u> Reinforces elimination of Muda (無駄, also ムダ), Reinforces Kaizen (改善, C.I.)

It is a foregone conclusion that most of the activities at this level will be directly tied to a downstream manager's goals and visible metrics. A

CIO who is assigned the cost goal of reducing overall spending by 10% of last year's budget should have a RYG visible metric with % Complete as a rollup from each of his supporting IT leader's budgetary efforts.

The monitoring iterations should be spaced far enough apart to allow for a cluster of course correction items such as regularly scheduled meetings, Genba walks, acting as the champion for a subordinate's Kaizen Events, and reassigning responsibility and accountability.

T.O.N.E. Pattern Examples

"What do you mean? How did we miss our sales target for the quarter?" In one Fortune 500 company, a director of a large business unit recalled this story:

> *"One particular quarter, each business unit missed on their sales related initiatives by anywhere from $.5M to $1M. In the grand scheme of things this iterative goal seemed of little consequence to each of the directors. Coming into the Operations Review meeting, the aggregate "miss" for the end of the upcoming quarter was around $10M. From the General Manager's point of view, this was unacceptable. In her mind, unless they quickly implemented some control measures, there was reason to believe the facility would miss its year-end goal by a whopping 5%. At this point (before Lean) the only resolutions offered were things like, "We will work to get this corrected," or "We just have to keep our eye on the ball a little better."*

Eventually, I ended up reporting to one of the directors. Each of her four managers had some share of a $6M reduction in raw material costs. We were to achieve the savings via a number of different purchasing vehicles with our respective suppliers. She set the group's success in motion by deploying a new strategy to monitor our progress. The first action the leader took was to set the monitoring iterations at one month intervals.

The new process was to:
1) Aggregate totals.
2) Analyze progress reports.
3) Find out what practices were working.
4) Aggregate best practices.
5) Triage any special cases, "Do you need my assistance?" "What can I do to help you to be more successful?"
6) Level set expectations for the next iteration. Readjust any unrealistically high or low goals (UCL / LCL).

These small batch sizes in the form of monthly checkpoints eliminated the element of chance in our success equation. During the year, our director sponsored a Kaizen Event where we reduced the time required to make raw materials available by shortening the quality inspection cycle. We started accepting more of the reliable materials on certification from the vendor and cleaned up miscues in the Receiving Department normally occurring at the end of the month. Several different departments and business units had to work together, map processes, eliminate waste, and brainstorm on how to track progress.

The interesting thing is that we were able to take this approach and use it as a template for any revenue related goal. The first department to utilize the new template was the IT department. We used it to streamline the budgetary process and drive down corporate allocation costs in areas of "waste." In a specific example, we invested in bandwidth across our LAN/WAN instead of the initial plan to implement LAN / WAN redundancy. This was based on the voice of the customer. At this particular juncture, there were no catastrophic failures driving the need for redundancy; however, there were monthly issues from scores of users reporting slowed response time while using applications, ad hoc queries and internet/intranet access. This led to the disassembly of our planned upgrades to the network and a reduction in recurring costs of over $100k for the year. This was not necessarily a great testimony to the normal process since these budgetary reductions were not a part of the responsibilities of management structure below the director level. It was the new Lean process' iterative monitoring that brought this hidden opportunity to the surface.

T.O.N.E. Review

I have often heard leaders say, "Early course corrections are essential for large ships and large organizations to avoid hitting the proverbial iceberg." This is certainly true of work in the T.O.N.E. pattern. Leaders and individuals holding key roles and responsibility for long-term initiatives should seek to implement Lean principles such as:

1) Visual graphs tracking Cost, Quality and Customer Satisfaction trends over some extended period of time.
2) Developing continuous feedback loops on key metrics to serve as an "Early Warning and Detection System" to initiate regular course corrections.
3) Supporting manager's visual boards. Metrics should "roll-up" to the leader's board to ensure continuity and accountability of goals.
4) Use standard work to triage problems at lower levels and provide support. A good model is:
    a. Aggregate totals.
    b. Analyze progress reports.
    c. Find out what functions are working and aggregate them into new best practices.
    d. Triage any special cases. Ask, "Do you need my assistance?" and "What can I do to help you to be more successful?"
    e. Level set expectations for the next iteration. Readjust any unrealistically high or low goals (UCL / LCL)

Chapter 6: Case Studies and Tools

Each case study follows the pattern of creating a Lean Culture shown in Figure 6.1. The examples use the Plan, Do, Check and Act cycle as a way to segregate activities.

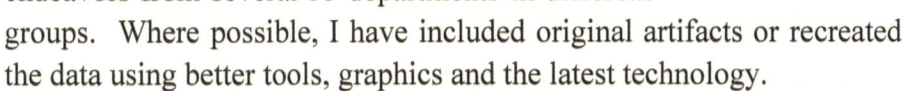

Many of these examples are taken from actual endeavors from several IT departments in different groups. Where possible, I have included original artifacts or recreated the data using better tools, graphics and the latest technology.

Many lean practitioners have some "one-off" version of this concept. Most experts hold to the basic premise that our organizations must practice a continuous cycle of activities in order to create a lean culture and not just the flavor of the month. These keys give the Pattern Matching Methodology its true power in that the cycle can be repeated in each pattern.

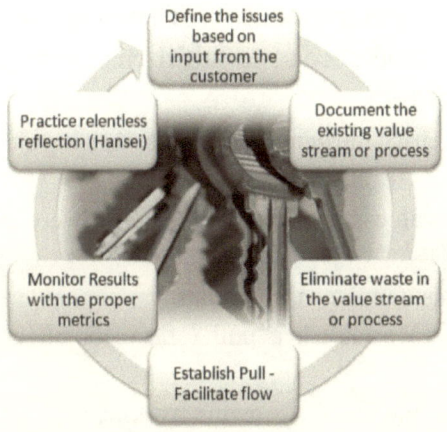

Figure 6.1 – Cycle of Creating a Lean Culture

R.P.M. Pattern Case Study

Case

Stated Issue: The hardware support group recognized the need to master the basic WCIT principle of Customer Service. The group decided to spend a portion of their software budget on queue / ticket management software after the survey review and initial discussion with a cross-section of the user community. Analysis of their specific situation revealed that the majority of their complaints came from change requests (Move my PC or phone, Change my configuration, or Add a new PC/Phone) instead of trouble tickets (my PC won't start, my phone is making a strange noise).

To validate this, the hardware support team decided against sending out another survey, but opted to take a "Gallup Poll" instead. While making their regular customer calls over the course of the next two weeks, they asked a series of carefully worded questions that got to the heart of the matter.

Plan

The evidence was conclusive; the hardware group was able to validate their data concerning change requests. They implemented the following plan to shorten the time needed to fulfill change requests and improve the quality of their work:

1) Establish metrics designed to move the needle for the customer. Establish goals for the metrics.
2) Complete a value stream map of the hardware change management process.
3) Eliminate waste in the value stream.
4) Find a way to establish a pull / flow for all categories of trouble tickets.
5) Review work and metrics – is this working?
6) Plan the next improvement cycle – up the ante!

Do

The customers informed the hardware group of their two most important issues:
- Getting scheduled requests completed on time, and
- Completing the requests correctly the first time.

With this in mind, they selected the following metrics to monitor progress with respect to the voice of the customer. The as-is state and goal for each of the metrics is annotated in parenthesis.
1) % of requests completed on-time (AS-IS = 17%, Goal = 90%)
2) % of requests correct the 1st time (AS-IS = 35%, Goal = 90%)

The group also decided to establish and set expectations for requests. They agreed with the customer's request for a 2-week lead time on all requests. This allowed the hardware group to accurately predict the work flow on a consistent basis. The following charts were developed to support the new metrics and goals.

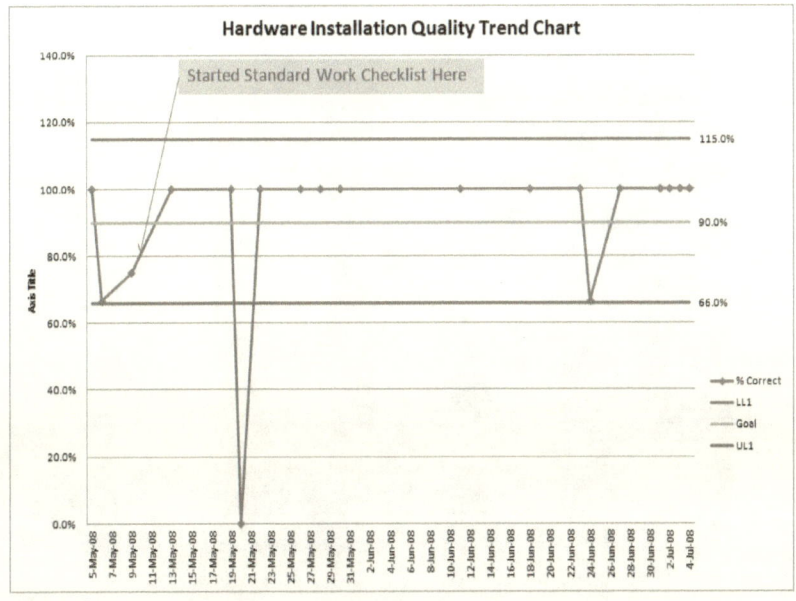

Figure 6.2 – Hardware Quality Control Chart

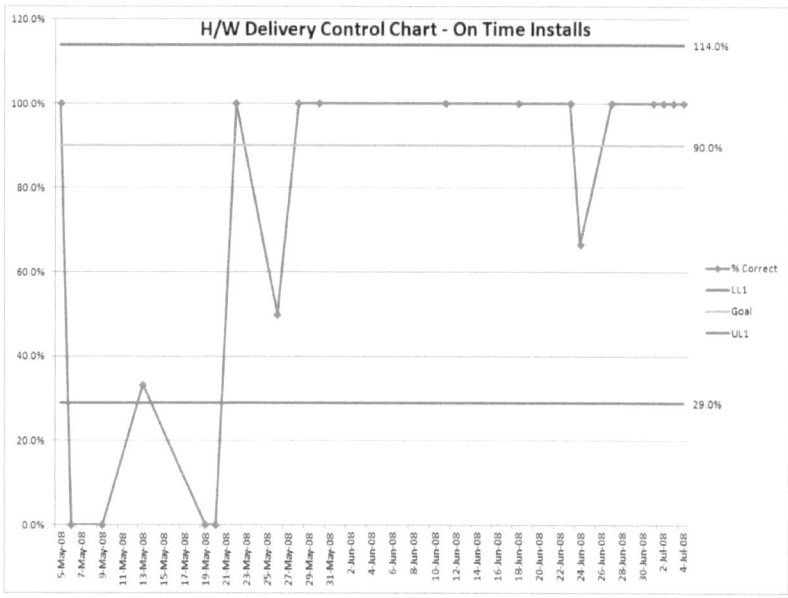

Figure 6.3 – Hardware Delivery Control Chart

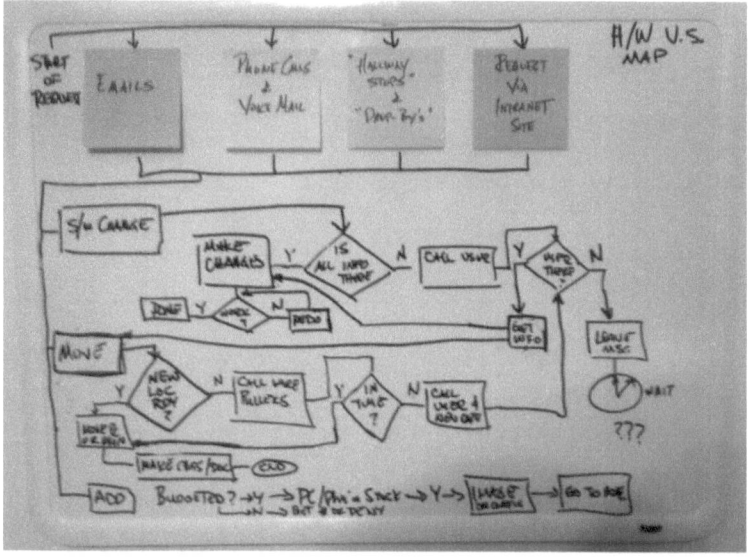

Figure 6.4 – Initial Value Stream Map for the Hardware Process

Process Description	Value Added?	Waste Category	Can this be Eliminated?	Time Required To Do Task (HRS)	Notes
Email Reqest	N	Processing	Y	0.167	Only one of these can be value added; the others are waste. Should be only one way to make a request
Phone Request	N	Processing	Y	0.25	
Hallway Request	N	Processing	Y	0.25	
Intranet Request	Y			0.167	
Review Request for Completeness	N	Processing	Y	0.25	We should have a fool-proof way to ensure the request has all of the needed info
Call customer to clarify info in request	N	Processing	Y	0.25	
Leave message / send email to customer	N	Processing	Y	0.25	
Waiting for customer to return call	N	Processing	Y	4	
Checking out new move location	??	Processing		0.75	Only if a new location is being created from an area that was not previously an office location. All office locations should be already configured
Prep new area (pull cable, install jacks)	??	Processing		8	

Figure 6.5 - Process Waste Elimination Worksheet (1ST page only)

Check

The team's next customer satisfaction survey conducted at the beginning of the 3rd quarter was glowing – nearly all of the participants noticed a marked improvement in the level of service. Along the way, it was interesting to note that many of the customers knew where our department stood with respect to both the Quality and OTD metrics!

Act

Flow and Pull –

The first issue the hardware team attempted to address was the uneven flow of the process. After a month of monitoring their main issues, they created a Heijunka Board to better manage customer demand. By the time the next scheduled the customer survey rolled around, the group fine-tuned their Heijunka Board and mastered the process of meeting and reviewing the results. A sample of the first draft appears in Figure 6.6. The idea was threefold:
1) Create a more even workflow during the week.
2) Exceed customer expectations with early installations, where feasible.
3) Avoid wasted motion; schedule work so that all installs/fixes for PCs in close proximity to one another (in the same building) are scheduled on the same day instead of attempting to cover multiple locations per day.

This new strategy eliminated wasted motion (travelling between buildings) in the process. We were able to exceed our customer's expectations by completing more installs on schedule. The dip in On-Time Delivery (OTD) in the month of June (see Figure 6.2) was due to a re-negotiated date with the customer in order to level the workload. The group later decided against re-negotiating work past expected install dates just to satisfy flow; in other words, they did not want to violate one Lean objective (on-time delivery) for another (level-loading).

These changes set the group in motion for making future improvements such as JIT hardware delivery and setting up min/max Kanbans. Additionally, the group asked management to convince the vendor to burn a standard image on new PCs. This saved additional setup and processing time. The following year, the hardware team budgeted additional dollars to improve the infrastructure by contracting a third party vendor to wire and document all known and anticipated office locations. Lastly, the group worked with management to reduce waste by developing an entirely different process for dealing with unbudgeted requests. These requests involved many additional steps outside the hardware group's scope of influence. These unbudgeted requests could flow through normal channels once they made it through the out-of-cycle financial and purchasing processes.

These additional Lean implementation changes and kaizen efforts occurred over the span of two years. The group routinely held weekly standup sessions where they reviewed the results of their metrics and eliminated any barriers preventing them from meeting their goals. Whenever there was a problem, they would alert the sensei. He would immediately facilitate a meeting and encourage the use of Lean tools (5-Whys, A3, etc.) to get to the root cause of issues related to the process.

The team also noticed that the Quality and OTD charts were nearly identical in places. Although several people postulated that there might be a connection, the group ultimately came to the conclusion that the improvements in quality were related to the adoption of standardized work. The improvements in OTD were directly related to flow activities connected with the Heijunka Board and work location consolidation.

While the group made significant changes and lowered both hardware and labor costs, the true success was the creation of a new Lean culture within the department.

JUNE

SUNDAY	MONDAY	TUESDAY	WEDNESDAY	THURSDAY	FRIDAY	SATURDAY
	1	2	3	4	5	
	8	9	10	11 1 sched	12	
	15	16	17	18 1 sched	19	
	22	23 3 sched	24 3 sched	25	26	27 2 sched 1 vacation next week
	29	30				

Mon: 15	Tue: 16	Wed: 17	Thu: 18	Fri: 19
8 Do Bldg1 install scheduled for 18"	8 Do Bldg 3 install #1 Scheduled for 23"	8 Do Bldg 35 install scheduled for 24"	8 Do Bldg 5 install scheduled for 24"	8 Do Bldg 5 install scheduled for 23"
9	9	9	9	9
10	10	10	10	10
11	11	11	11	11
12	12	12	12	12
1	1	1	1	1
2	2 Do Bldg 3 install #2 Scheduled for 24"	2	2 Do Bldg 5 install scheduled for 23"	2
3	3	3	3	3
4	4	4	4	4
5	5	5	5	5
6	6	6	6	6
Evening	Evening	Evening	Evening	Evening

Figure 6.6 – PC Support Group Heijunka Board

P.I.C. Pattern Case Study

I previously mentioned that a number of books, periodicals and methodologies show how Lean can be applied to software and product development. Consequently, I will use a different IT discipline's case study as an example of a group utilizing the P.I.C pattern to establish Lean.

Case

Stated Issue: The newly formed Business Analysis / Project Management group noticed an all too familiar pattern; none of their projects were coming in on time. More importantly, there seemed to be a disconnect with stakeholders concerning expected outcomes of these projects. Often, the customers would end up with something less than they expected, with features they didn't need or want, and with applications and tools requiring extensive and unbudgeted training. The lack of understanding of scope and functionality often led to budget overruns. Often, the IT department had to extend service contracts with vendors for implementation services. It became imperative to come up with a way to standardize the process and reduce the labor force and budgetary impact to the business. Every year there seemed to be large-scale, corporate-mandated and/or internally generated application development projects slated for implementation. Occasionally, there were also one or two IT generated upgrades (like Microsoft Office releases).

Initially, individuals responsible for infusing technology into the company were skeptical about trying to adopt Lean into their groups. No one saw how the practices could possibly be instituted. A fair amount of cognitive dissonance hung in the air regarding the abolishment of standard project management practices. The solution is not to abandon all conventional wisdom and disciplines, but rather to eliminate the waste in this pattern, to shorten the batch cycles and add governance (standardized work) in order to affect quality and outcomes.

The key stakeholders in the group reluctantly agreed to proceed with a Lean journey.

Plan

The plan was to follow the same Lean Culture Cycle template (Figure 6.1) used by other groups in the IT organization:
1) Establish metrics to move the needle for the customer. Establish goals for the metrics.
2) Do a value stream map of the business analysis / project management process
3) Eliminate waste in the value stream.
4) Find a way to establish a pull / flow with the requests.
5) Review work and metrics – is this working?
6) Plan the next improvement cycle – up the ante!

Do

The customer's reaction to our question about what was most important was predictable. This group of stakeholders felt uncomfortable with our ability to predict end / delivery dates for large-scale projects. Although some budget overruns within reason were to be expected, they still wanted to see the group exercise better controls in both areas. After all, the converse was seldom if ever true – that is, the likelihood of a project coming in ahead of time and under budget.

The most unacceptable outcome of the process was the quality of the products we delivered. Too often key functionality was missing or key features were misunderstood. The customer felt that the IT professionals never really understood their work. They felt that the COTS software we recommended was fraught with features that were not needed, not applicable or unexpected. The one universal truth all IT people have experienced is that users don't like being surprised. It is almost impossible to get 100% buy-in of a new application or platform, but most of the "user-ware" existing on PCs in the form of spreadsheets and Access Databases is a result of the user giving up on the available functionality of the current software application.

Taking all this into consideration, the group listed the main goals as:
1) Capturing concise, exact requirements 100% of the time.
2) IT takes responsibility for communicating the scope, impact, and changes of any large scale upgrade.
3) IT develops processes to improve forecasting accurately, report status and risk, and budget containment.
4) IT develops a Service Level Agreement with the customer outlining standard expectations such as frequency of communication, status updates, testing and training and support.
5) IT eliminates internal scope creep on software development.

Gathering Requirements

The group brainstormed and came up with a standardized work process to gather requirements for the different types of projects. The model is depicted in Figure 6.7. They were able to ensure completeness and predictability in nearly all time estimates, regardless of the analyst in charge, by standardizing the requirement gathering process.

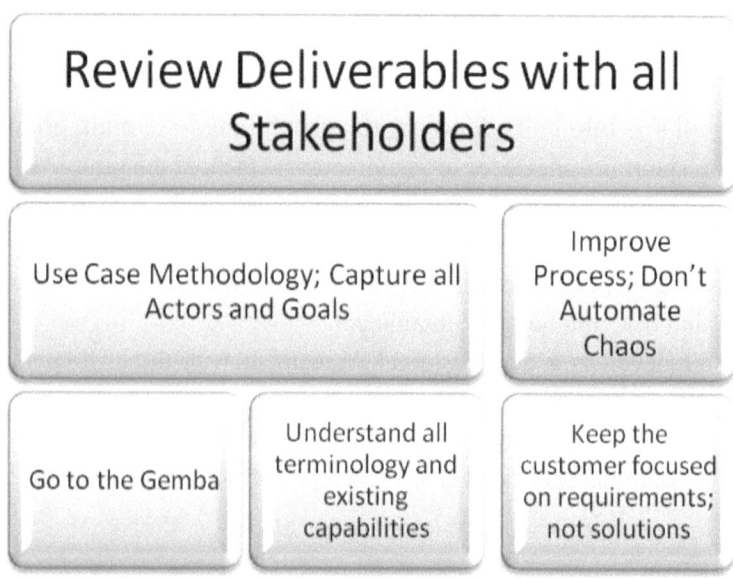

Figure 6.7 – Requirements Gathering Process

Communications and Governance for Establishing and Changing Scope and Impact

The group implemented additional governance and established Service Level Agreements communicating Scope and Impact. They also addressed the need for governance regarding projects requiring signoff from multiple business units. This turned into a Group (multiple locations with the company) effort since everyone saw these disciplines as baseline World Class IT capabilities with widespread adoptability and scalability. In this pattern, governance represents the concept of implementing standardized work procedures.

This took quite some time. It required multiple meetings with stakeholders and executives but proved to be a crucial step and a major tool in preventing waste in the form of waiting and overproduction.

These documents were staged in a central repository for all project initiators to review, fill out and sign off.

Visual Indicators and Metrics

Following the newly established methodology for creating a Lean culture, the group began by documenting the as-is process. The following artifact (Figure 6.8) shows that they kept almost none of the existing methodologies. It was a good time to overhaul the proverbial engine since the business analyst positions were new to the IT organization. The group admittedly approached the VSM exercise at the macro level, but even after this initial effort they realized the massive amount of waste in their processes related to scheduling meetings, attending meetings, waiting on items, re-meeting, etc.

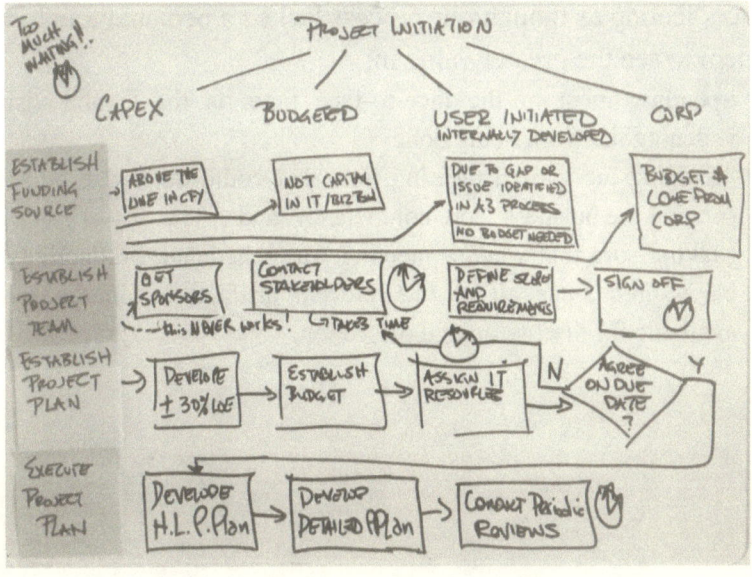

Figure 6.8 – Project Management VSM

The group calculated (using a time study) that nearly 65 – 70% of the time it took to define and execute a project was spent waiting on deliverables and sitting in meetings. The group immediately revamped the entire process. They hosted a kick-off meeting to explain their newest project and unveiled the new process. The process involved a series of comprehensive electronic forms staged in a central repository, a new project status intranet site, and document workflow software. Periodically, the group would still need to schedule a meeting; however, the amount of time spent waiting, following up, and scheduling meetings dropped significantly.

This was a huge cultural change! We initially faced a fair amount of resistance from many organizations. In the earlier stages, the business analyst would stave off the resistance by offering to scribe and transcribe the requirements into the electronic forms. Later, we found key individuals in each business unit who were comfortable working with electronic forms and technology. While we did not want the customers feeling as though our process lacked a personal touch, we did want them to see the greater value in:

1) Spending most of the face-to-face time on the Genba visits and watching the work being done,
2) Avoiding the wasted meeting time that could otherwise be spent on running the business' core competency, and
3) Making sure that when a meeting was scheduled, productivity could be assured since all of the relevant artifacts to do the job were available for discussion and dissection.

Process Description	Value Added?	Waste Category	Can this be Eliminated?	Time Required To Do Task (HRS)	Notes
Email Reqest	N	Processing	Y	0.167	Only one of these can be value added; the others are waste. Should be only one way to make a request
Phone Request	N	Processing	Y	0.25	
Hallway Request	N	Processing	Y	0.25	
Intranet Request	Y			0.167	
Review Request for Completeness	N	Processing	Y	0.25	We should have a fool-proof way to ensure the request has all of the needed info
Call customer to clarify info in request	N	Processing	Y	0.25	
Leave message / send email to customer	N	Processing	Y	0.25	
Waiting for customer to return call	N	Processing	Y	4	
Checking out new move location	??	Processing		0.75	Only if a new location is being created from an area that was not previously an office location. All office locations should be already configured
Prep new area (pull cable, install jacks)	??	Processing		8	

Figure 6.9 – Project Management / Business Analyst Process Elimination Sheet

Lastly, the group used an entirely new set of visuals to track projects based on the customer's input. The combination of the new metrics allowed early detection of issues and a more comprehensive "storyboard." This replaced the typical 10-page project plan with items check off and an overall percent complete number. The customer immediately understood where the issues were, when the project was headed in the wrong direction and which specific critical tasks were threatening to put the entire project at risk.

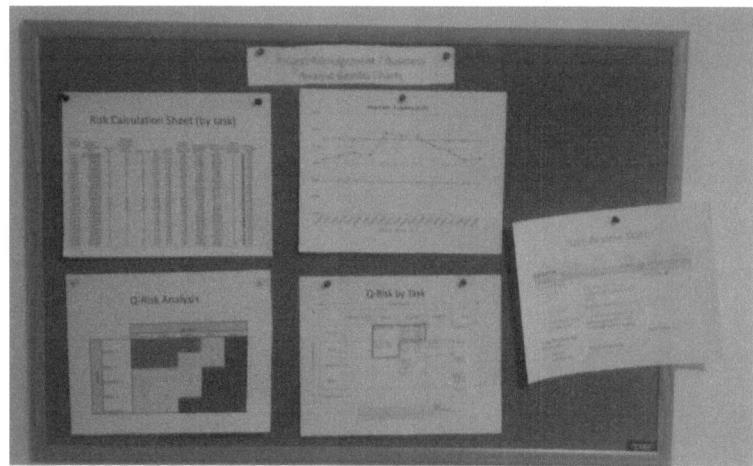

Figure 6.10 – Project Management / Business Analyst Genba Charts

QRisk Analysis		Likelihood				
		No Chance - 5	Unlikely - 4	Possible - 3	Likely - 2	Certain - 1
Severity	Severe - 5	0	1	4	0	1
	Major - 4	0	2	3	2	1
	Moderate - 3	0	0	3	0	2
	Minor - 2	0	0	1	0	3
	Insignificant - 1	0	0	1	0	1

Figure 6.11 – Qualitative Risk Analysis Visual Chart

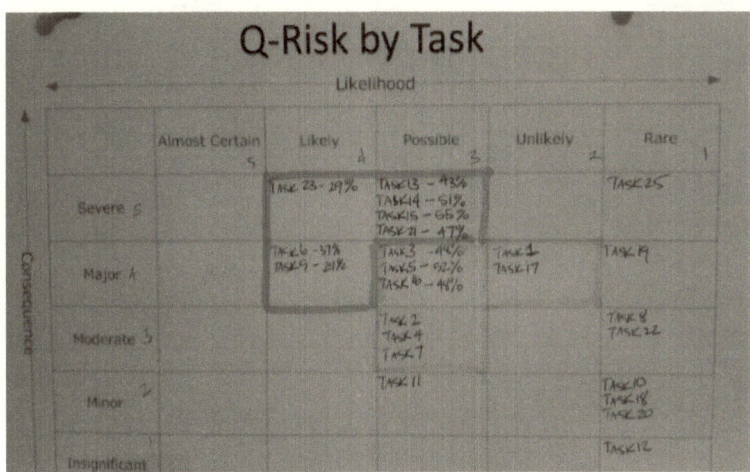

Figure 6.12 – Qualitative Risk Analysis by Task

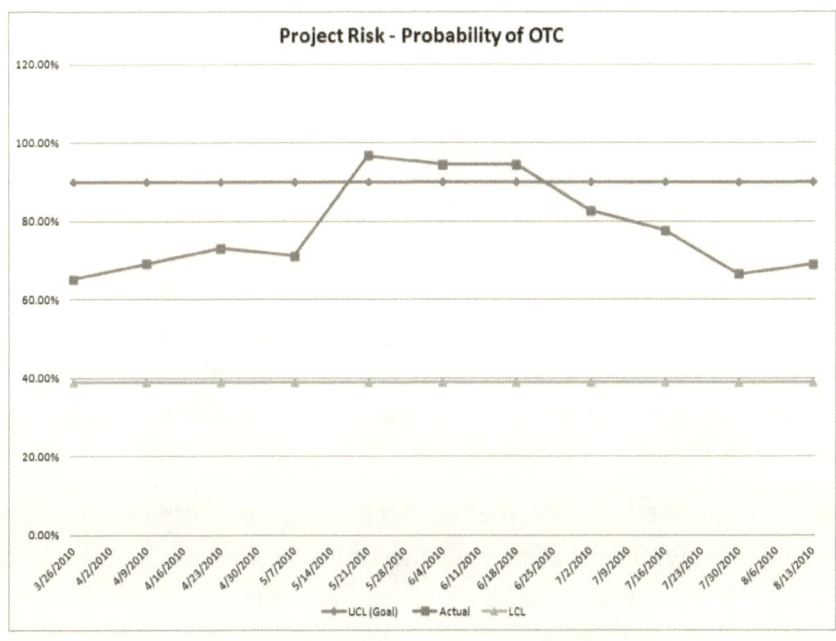

Figure 6.13 – Probability of Success Control Chart

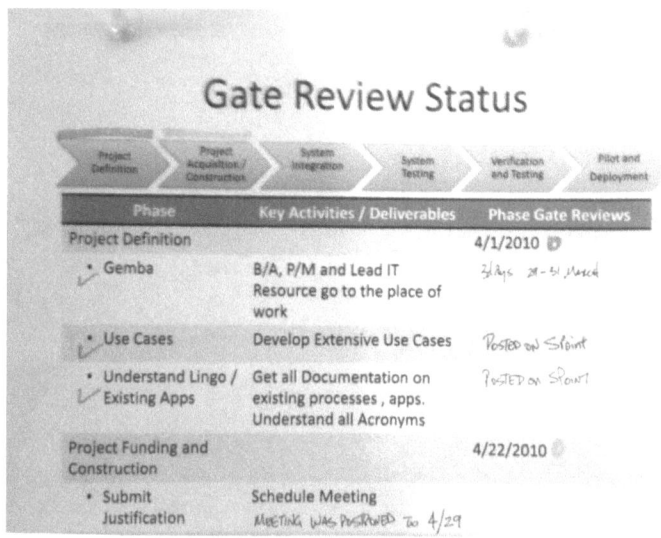

Figure 6.14 – Gate Review

Check

The results in this area were almost too good to be true. In the past, many customers were very frustrated because of the seemingly endless amount of time needed to support and see a project to completion. A frequent question posed to IT was, "Can you tell me how much time my people are going to have to spend on this?" The answer was almost always the same; "We can't tell you exactly how much, but it will be…" With the new focus on shortening the "manufacturing" cycle, IT could tell the customer exactly when and how long meetings would last and provide them with an approximate time frame for putting together the project artifacts and requirements.

Almost immediately after the changes went into effect, customers noticed the difference and offered the following feedback (taken from the quarterly customer surveys) after the first three projects were past the project definition stage:

> *"I know exactly how my project is going and what the issues are. I love the weekly updates."*
>
> *"It feels like they have nothing to hide now..."*
>
> *"I have a greater degree of comfort that issues will be resolved. Any time I want status, I can just walk by the board without even getting someone from IT involved. It's great!"*

The group found the following tools worked well for these types of responsibilities:
1. Establishing governance,
2. Using R/Y/G indicators,
3. Monitoring progress using gates, and
4. Implementing quantitative risk monitoring.

This early warning detection system provided them with an ANDON cord to pull as soon as they noticed the project getting out of control.

Adding this to the gate review process reduced the batch sizes and set the project up for success.

Act

The group undertook and developed an A3 they felt would help eliminate the ambiguity associated with understanding workflow and expectations. Typically, the team never knew how many projects were in the pipeline for any one year, but with a streamlined process, they could predict production outputs based on current staffing levels. By knowing this, they were able to give the leadership team some visibility and flexibility as to what options they had towards prioritizing and staffing for the year.

The budgeting process was changed to include anticipated projects. Now the team looked at the existing backlog plus the upcoming projects and was able to accurately predict what could and could not be finished that year. After allowing for a few high profile "emergency" projects, IT was able to take forward a list of the customer demands. The leadership team's job was to help validate and prioritize the list before the beginning of the new fiscal year.

As a follow up, the group came up with Takt time calculations for their work, but since a "unit" was fairly heterogeneous (a document, a meeting, a section of code, a phone call to a vendor, reviewing technical specifications, etc.) and since every project called for a slightly different recipe, they abandoned its use. It did not increase the overall throughput nor did the customer seem to see any value in the metric.

These tools worked especially well with the Agile projects in the product / software development group; however, it had the greatest impact on the corporate, consulting and capital projects. These changes took place over the course of two years; instituting governance was the biggest consumer of time in the entire process. In hindsight, we realized most of this came about by cajoling, negotiating and, in some cases, by brute force. A marketing campaign much like the one executed by the hardware group would have facilitated a much smoother transition for the customers receiving services in this work pattern.

T.O.N.E. Pattern Case Study

Case

Stated Issue: The Supply Chain Director's balanced scorecard contained no less than 15 initiatives related to improving the sales and cash position of the company. There were a handful of things that could only be accomplished through cooperating with the other business leaders, but there were two things that stood out related specifically to Supply Chain; reducing the spend on materials by $6M and holding the overall Material Price Index (MPI) under 1.0.

Plan

This is a typical setup; this director was given a broad goal. Knowing the time between checkpoints was too infrequent, the leader seized on the opportunity to champion Lean in the organization.

The leader immediately kicked off a comprehensive campaign in the organization to begin creating a Lean culture. Partnering with the IT leadership team would prove to be critical since the Supply Chain organization is usually the biggest consumer of IT resources in most manufacturing businesses. The two leaders immediately realized the potential for a disconnect between their two organizations.

The executives immediately noticed that the voice of the customer was missing with respect to these two metrics. The Supply Chain organization viewed IT as a critical partner in assisting them in converting data into information; however, the ERP system did not aggregate this information in a way to allow the managers to frequently monitor their progress. After listing a number of other concerns, the two executives decided to make these goals a joint effort.

To eliminate large batch sizes that normally sabotaged year-long goals like these, the plan was to:
- Create a visual signal showing progress on a monthly basis.
- Create a process to kick off a standardized list of Lean problem resolution activities in the event of a signal indicating an abnormal condition in with the metric(s).
- Ensure the visual signals on the executive's board were directly tied to the visual signals in their direct reports areas.

Do

The following visual charts were developed for the Leadership Team. The leaders preferred to have hand-written, visual indicators hanging outside their offices. The admin would update the status a minimum of twice a month. At any time, the leadership team could go to one of the support areas and drill down into either of the two metrics to see a weekly control chart by area (Direct / Indirect Materials, Modernization, etc.).

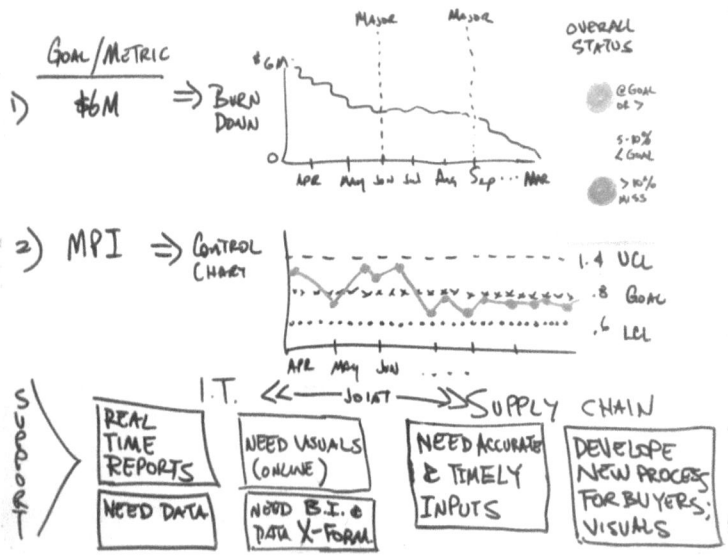

Figure 6.15 – Initial IT & Supply Chain Executive Planning Session

In the event the metrics started to veer away from the goals, the leaders agreed to the standardized work shown in figure 6.16 to remediate the situation. The idea was to call a quick "all-hands-on-deck" meeting with all parties involved. This "ANDON cord" got pulled anytime the monthly metrics strayed out of the control area. Everyone knew what to expect when this happened and it would always follow the same format:

1) Send out notifications for the meeting. We made an effort to schedule it during the day, but on such short notice, we would typically have the first meeting at lunch or after hours.
2) IT starts the Totals Aggregation. This usually meant they had to retrieve the data from the backups from the day the reports were run. After the "Mini B.I." solution was in place, Supply Chain simply ran the reports using the appropriate views. This was different than the normal reports in that the summary and detail data were both available for analysis.
3) The next day, SC managers and subject matter experts gathered together to analyze the data. The leadership team was not involved at this time since some of the issues might have been related to entry errors, late data entries, or process related problems. If the group determined a real problem existed, they scheduled a meeting to present a 5-Why analysis and aggregated Root Cause resolution(s) to the executives.
4) Sometimes, the issues were related to special cases which required executive intervention. These were annotated in the presentation and the leaders were given a list of suggested actions.
5) After the leadership team determined that satisfactory Lean controls were in place and the changes to the process had been poka-yoked (error-proofed), they would level set expectations. This meant ensuring that Lean principles were still being followed and that the entire organization understood the collective commitment to the Lean journey.
6) In the same meeting, the goals were amended, if necessary. In one particular meeting, the reason for amending the goal was due to the fact that the market price for one of the raw materials had nearly doubled in the past month. There was little that could be done except to wait for the price to drop again. In the meantime, we changed the parameters of the goal and alerted the appropriate business leaders of the change and its ramifications to the goal well in advance of the end of the year.

Organization			Standardized Worksheet							
Standardized Work Sheet			Operation Name	Metric Recalibration	End Product	STABILIZED MPI / COST SAVING				
Takt Time	3 Days		Work Group	IT and SC	Lead				Page	1 of 1
Cycle Time	n/a			Time		Manual	Automatic	Waiting		
Step	Step Description		Manual	Auto	.5d	1d	1.5d	2.0d	2.5d	3.0d
1	Schedule Meetings		15m							
2	Have I.T. aggregate totals		1hr							
3	Analyze Progress Reports		.5d							
4	Find out what's working		.5d							
5	Aggregate Best Practices		.5d							
6	Triage special cases		1d							

Figure 6.16 – Standardized Worksheet for "Crisis Intervention"

Check

This led to some obvious improvements such as standardized processes related to data entry, vendor management, contract administration and even consolidated shipping. While the support teams worked on the week-to-week issues, the leaders were able to keep a firm grip on the helm of the ship. Making small course corrections ensured there would be less of a chance the group would be in danger goal of missing its goal. At the "11th hour" of the year, there is little that could be done to salvage a company's position.

Act

This approach quickly spread to other areas in the leadership ranks. The previous strategy was to take great care in selecting competent managers and hope for the best. Now the executive were taking an active role in moving the Lean initiatives forward as well as keeping up with mission critical goals. These periodic checkpoint meetings and problem resolution sessions, if needed, only took an additional 2 – 3 hours out of the leader's schedule each month. These meetings were never to be delegated except in the case of an emergency.

Chapter 7: Putting IT all together...Are We Winning?

Our Journey in review

Deploying these methodology patterns allows us to cover a lot of ground with typical IT roles. Consider this matrix of responsibilities matched with patterns:

IT Function	Suggested Pattern
Help Desk	RPM
Server Administration	RPM or PIC
PC Support	RPM
Application Support and Administration	RPM
SOX Administration	RPM
Application Development	PIC or RPM
Network Administration and Security	PIC
Infrastructure Administration, Design, Installation and Support	PIC
Business Analysis / Customer Requirements	PIC
Project Management and Portfolio Selection	PIC or TONE
Planning and Strategic	TONE
Executive Leadership	TONE

It is important not to put the adoption of these tools ahead of the steps responsible for creating the culture. The organization must understand that the primary opportunity Lean affords is to shorten the work cycle. Other process improvement methodologies focus on the producing more product, usually without respect to customer demand. A few others focus on improving quality. Almost none focus on fostering a culture where all workers are encouraged to use their creativity to solve problems and improve processes. Lean focuses on all of the above and more.

With Lean, the focus is on eliminating steps, motion, overproduction and other wastes from the process. Eliminations of this type will naturally increase our ability to produce more, but these should take a backseat to the demand pace dictated by the voice of the customer. Again, we do not overproduce or provide features and/or capabilities that are not important to the customer. This focus also builds in quality and eliminates another waste, rework.

When we focus on producing more, we end up with:
1) Ever expanding work queues,
2) Large batch cycles that are hard to track and monitor,
3) Repeating instances of poor quality,
4) Endless justification requests for more headcount, and
5) Frustrated customers / users.

Consequently, our business units and corporate structures never see us as business partners who should be included in strategic discussions. Instead, we are viewed as a necessary evil.

Once again, culture is created by understanding the customer's hierarchy of needs from their respective IT departments (see Figure 7.1). This way IT can first understand how to define and provide value, map processes with the intent of exposing and eliminating waste, understand how to create pull and flow within each work pattern, and pursue Kaizen.

IT Customer Hierarchy of Needs

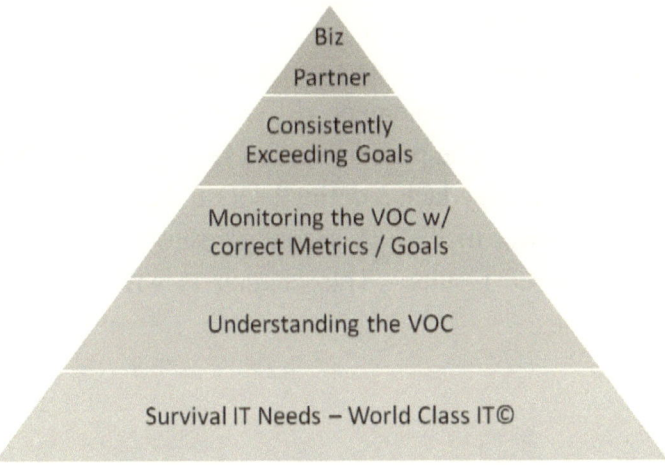

Figure 7.1 – What the Customer Needs from IT

It takes time to implement Lean. It takes patience and persistence to change a culture. It is absolutely necessary to be willing to change our patterns of thinking and the way we resolve issues. Approaching Lean must be done differently than the approach used to implement other methodologies, solutions, and tools.

While researching ways to submit estimates to users and how to effectively manage changes in technology, I read an internet article decrying repetitive efforts by stating, "Remember, the definition of insanity is doing the same thing over and over again and expecting different results." The article then incorrectly attributed the quote to Albert Einstein. This thinking permeates business culture in Western Society and is not universally applicable to every situation. Lean demands the persistence and patience that goes hand-in-hand with leading permanent change. In the book, "The Lean Startup," the author defines a "pivot" as a change in strategy without a change in vision. (Reis, 2011) The vision to implement Lean requires a pivot in our strategy from one of fast implementation to one of progress through persistence. Tactical pivots can be made quickly. The idea is to persevere.

I have been privileged to be a part of transformational IT departments that took the time to implement different parts of the collective strategies in this book. The primary reason IT organizations do not take advantage of Lean is they lack a defined set of blueprints with which to build the "Lean Tower of IT." A secondary reason is that most feel it takes too long to see results. This chapter is written specifically to address these issues.

When is the Journey Complete?

Lean is a journey. Toyota has been implementing Lean for over 40 years and they report that they are still not finished. This fact is sometimes confused with wondering when teams should stop seeking process improvement or how long they should wait before putting another process in place. My favorite response is, "When making another change would put you in *zugszwang*."

Zugszwang is a German term meaning, "compulsion to move." Its common colloquial use is in the world of chess and it is a term best illustrated by a game between Marshall and Capablanca in 1918 (Queen's Gambit Declined). In the opening moves of the game, Capablanca made an early pawn sacrifice. This was part of an overall strategy to lure Marshall into a position where all of his pieces were tied up or pinned down. Marshall was placed in *"zugszwang"*; that is, *any move he made would put him in a worse position than he was already in.* ** (Chernev, 1992)

Anytime there is a change in business, it is necessary to reexamine the process, but be careful about changing a process already free of waste and non-value added steps. A common trap IT professionals fall into is attempting to automate everything or to turn everything into an electronic process. In some cases, it makes absolute sense to do so, especially if it error-proofs the process. However, the proclivity to do this is often premature and leads to the trap of *automating chaos*.

It is more important to know when to re-start improvement efforts. Many times, organizations relentlessly pursue improving their processes; however, they neglect to re-evaluate them once the product, goal, organization or business goes through important changes.

**- By the end of the 34th move, Capablanca was in the envious position of having doubled Rooks on the seventh rank! Marshall resigned after the 39th move.

Figure 7.2 – Zugszwang; any move white makes (clear pieces) worsens the situation.

Chapter 8: The Voice of the Customer…Revisited

I handed a draft copy of this book to one of my IT colleagues for review. After reading it, he asked me a really good question. "So I get the fact that I cannot approach Lean IT as just another process or methodology. And I can see from what I've read that after we  have stabilized our foundation, I should start by understanding what my internal customers want. What I want to know is; who do I start with? Am I to track metrics and goals for every separate department and customer?"

After pondering this a bit, I elected to pull bits and pieces of information dispersed throughout the book and put them together in one focused chapter to maximize clarity.

It would take no great amount of cajoling to get IT executives, managers and individual contributors to admit that there will always be a subset of users who put into practice the old adage, "The squeaky wheel gets the grease." We could all cite example after example where a particular user or a business unit within the company pushed their agenda at the expense of the rest of the corporation. But if IT is to truly move past being considered an ancillary department, we must demonstrate our business acumen as well as our technical chops. Too often, the IT department reports up through some other business unit like Finance or Supply Chain instead of having a seat at the leadership table and being considered an integral business partner. We must be able to clear a path to the most important issues in our respective businesses. We must be able to recognize the voice of our key customers above the noise of the "squeaky wheel."

In the 2011, Harvard Business Review published a compilation of articles in the e-book titled, "Aligning Technology with Strategy". In the article titled, "Six IT Decisions Your IT People Shouldn't Make," authors Jeanne W. Ross and Peter Weill make the following observations:

> *"The companies that manage their IT investments most successfully generate returns that are as much as 40% higher than those of their competitors."*
>
> *"Decide which IT initiatives will further your strategy – and fund only those. You'll avoid burying your IT department in irrelevant projects."*
> (Weill, 2011)

One of the biggest advantages to implementing Lean IT is that it helps the IT organization focus on the right voice, the right issues, the right processes, and the right investments.

Every corporation or business has a primary product or service that generates a lion's share of the revenue and business for the company. It is the company's core competency. It is what gives the company its competitive advantage and is, in most cases, the business unit that keeps the recipe to the "secret sauce." It is the one department in the company that will not be outsourced. IT organizations must recognize this in order to navigate through the complex labyrinth of requests for time, service and investment.

During my time at Sprint, the future of the company was mortgaged on a thriving long distance service offering. Long distance…my how things have changed! Most of our technological advantage was tied up in our long distance infrastructure. At the time, there was a separate IT organization dedicated to servicing the needs of the long distance network's complex interlocking of DMS-250/300 switches. Within each discipline of IT, we offered services to provide care, feeding and forward vision to maintain the revenue stream. We provided support in the form of a 24x7 Network Operations Center, Call Processing and

Provisioning systems that fed the billing systems and technical consultants who were tasked with keeping pace with the latest technology and engineering breakthroughs – all in order to gain a competitive advantage. Similarly, in the manufacturing sector, the Production / Operations department is usually the central hub and life force of the company. The department that most affects their success is the Supply Chain organization. Failing to focus on the needs of these customers will result in an IT organization that is generally viewed as ineffective in helping to move the mission forward.

There are obvious choices an executive can make to keep the focus on the right disciplines. A good example is an executive's decision to prioritize the investment in an ERP system over the upgrade of the HR system. The analysis may show that the ERP investment's rate of return, increase in profitability and labor reductions give them a competitive advantage. The much more difficult decision is what to do with the request to adopt Business Intelligence (BI), implement a service integration hub, portal technology or to add a shop-floor data network. Which of these requests is going to yield the greatest return on our investment? How can the decision be quantified or extrapolated?

While we can certainly make a case for spreading the wealth and resources around to other organizations supporting the core business, a look at where the IT department spends its time and resources should reflect an effort proportionate to where the business is focusing their energy and investment. This may seem obvious, but if that is true, why do we see article after article explaining how business leaders are trying to redirect the focus of the IT organization?

There will be obvious synergies in each area that stand out and are germane to the entire company when we look at the combined voice of all customers. There will be common threads pointing to some technology that should be implemented or improved. To ensure that we are truly partnering with the business, it should be clear from our metrics and goals that our focus is centered on improving the company's core competencies.

Many companies have instituted a Steering Committee chartered with prioritizing IT requests, projects and investments. If you work at a place like this, much of your work in this area may already be complete. In other corporations, senior leadership has hired a CIO or CTO to handle the responsibility of working with key leaders and business units – an approach that certainly saves time and money. Even if neither scenario fits the description of your company and you are the lone manager or director left to make these decisions, a sound strategy is to start by interviewing the leaders who support the company's main product or service offering. Round out your effort with surveying the rest of the company. It is always a good practice to step back and see if what you are purporting to be centerpiece of your organization's effort makes sense and is in line with the company's direction.

Chapter 9: KATA – Implementing a Lasting Lean Culture

You may be tempted to put down the book at this point, but don't do it…this is my big dismount. While this will be one of the shortest chapters in the book, it drives home a key point so… *persevere*!

Enter Gary Perkerwicz. Gary is one of the smartest guys I know. He has a storied background in application development and over the course of many years, has established himself as an invaluable repository of knowledge. He is an expert at what I call the Read-Research-Experiment-Recommend cycle. So I was not at all taken aback when Gary came into my office with his laptop in one hand and an armful of books in the other.

Gary "strongly suggested" that we lacked a key component in establishing a Lean Culture at our corporate IT office. By this point in my career, I had successfully overseen several Lean initiatives and I felt we had a tried and tested commodity. But any true Lean practitioner could not pass up the opportunity to practice the "relentless pursuit towards perfection" so; I relented.

As it turns out, Gary would constantly keep me on my toes and immersed in the literature of some of the best Lean thinkers of our time. One of the key concepts he brought back from a training class was the concept of KATA.

Wikipedia describes Kata as follows:

> **Kata** (形, or more traditionally, 型 (literally: "form") is a Japanese word describing detailed patterns of movements practiced either solo or in pairs. (Wikipedia, Kata, 2011)

This term is most commonly used to describe the over 100 different forms in karate. Studying a new kata helps to develop your ability to visualize attacks and appropriately respond to them. There is no possible way to exactly predict how an attacker will come at you and the idea is not to attempt to teach rote moves to combat each attack. *The goal is to be able to adapt to any given situation.* At once, I begin to see why this term was use to explain the creation of a Lean culture. The wiki goes on to explain how the term has evolved as it has found its way into Western vernacular:

> *More recently, kata has come to be used in English in a more general or figurative sense, referring to any basic form, routine, or pattern of behavior that is practiced to various levels of mastery.* (DeMenthe, 2011)

> *In Japanese language, kata (though written as* 方 *is a frequently-used suffix meaning "way of doing," with emphasis on the form and order of the process. Other meanings are "training method" and "formal exercise." The goal of a painter's practicing, for example, is to merge his consciousness with his brush; the potter's with his clay; the garden designer's with the materials of the garden. Once such mastery is achieved, the theory goes, the doing of a thing perfectly is as easy as thinking it.* (Ichijo, 2011)

> *One of the things that characterizes [sic] an organization's culture are [sic] its kata - its routines of thinking and practice.* (Ichijo, 2011) *Edgar Schein suggests an organization's culture helps it cope with its environment* (Schein, 2011), *and one meaning of kata is, "a way to keep two things in sync or harmony with one another." A task for leaders and managers is to create and maintain the organizational culture through consistent role modeling, teaching, and coaching, which is in many ways analogous to how martial-arts kata are taught.*

What are your organization's routines of thinking and practice? This is a very pertinent question. It helps to frame our discussion concerning the development of a Lean culture, or a KATA within our respective organizations.

In basketball, after a team masters the basics such as ball-handling, lay-ups, shooting, etc., any good coach will begin to teach his players a few plays related to offense, defense and press break. When learning to play chess, the novice learns how all the pieces move and immediately begins an immersion in strategies and other chess matches played by masters of the game. When learning to play a wind instrument, one must begin by learning the basic fingerings, breathing techniques, scales and patterns. Afterwards, most students begin to study the solos of their favorite artists and imitate their improvisations note-for-note. But in reality -

- Knowing basketball plays does not automatically make a team a winner.
- Knowing all of the chess openings, middle and end game strategies does not necessarily make one a good chess player; a computer can be programmed with all of this information.
- Mimicking one's favorite jazz artist and being able to duplicate their solos does not mean one possesses the ability to improvise. It means you have the ability to copy.

We observe different forms of kata throughout our lives. A basketball play almost never develops the way the coach draws it up in practice. One cannot predict the way another chess player will move in response to a given strategy. Buy a concert ticket and go see your favorite recording artist. After you take your seat and the concert begins, you will discover that although you recognize your favorite song (and may have even studied the transcription), what you hear being sung or played is slightly different. Each performance, solo or improvisation is subject to slight variations each time the song is performed. Memorizing the alphabet or the pronunciation of words does not make one a better conversationalist. The tools are there, but mastery dictates that we know how to use the tools. Players know how to play, not because they know the plays or the music, but because they understand the game and the song.

Gary's focus has always been on helping to create culture by getting the resources "playing" as soon as possible. I am sure by now Gary has helped to foster the equivalent of a "John Coltrane Lean practitioner."

Everyone has heard the Chinese proverb, "Give a man a fish and you feed him for a day. Teach a man to fish and you feed him for a lifetime." In our efforts to implement Lean in knowledge organizations, which are we doing?

In the Western world, we are masters at reverse engineering anything from an invention to a process. We know how to implement things. In our efforts to mimic Toyota's Lean manufacturing success, we end up teaching our people new processes rather than teaching them to solve problems.

In his book, "Toyota KATA," Mike Rother recalls an instance where a friend of his revisited a Toyota factory in Japan. On his first visit, he witnessed an assembly line where flow-racks were in use. This allowed the assembler to grab whatever part was needed to manufacture a number of different products. On his subsequent visit, the same line was using "kitting" to manufacture products. His friend was a little perturbed by this development and wanted to know if the right solution was to implement flow-racks or kitting. The production manager's answer in short was, "it depends." What he didn't know was that the number of products being developed on that line had doubled since his first visit and the flow-rack configuration was no longer the best solution. The visitor was focusing on the implementation process while the production manager's focus was operating in a culture that always develops the right solution for the problem. (Rother, 2010)

Should you conduct a 5-Why or do you draft an A3? It depends. Should you fill out a kaizen card or just make a change? It depends. What we want to develop in our organizations are thinkers and problem solvers. Without the right culture, we will end up with "Lean doers." This is why Lean's biggest waste of all is defined as untapped human potential.

If we focus on process and implementation, all we will end up with is the flavor of the month. You will know this is what you have created because it is unsustainable. After the dust settles, *any change a leader has to drive is a process – if the people drive it, it's a culture*.

There are some notable differences in our thought patterns versus the Eastern way of thinking. The book, "Geography of Thought," lists some differences which might explain our natural bent:

- *Patterns of attention and perception, with Easterners attending more to environments and Westerners attending more to objects, and Easterners being more likely to detect relationships among events than Westerners.*
- *Basic assumptions about the composition of the world, with Easterners seeing substances where Westerners see objects.*
- *Beliefs about controllability of the environment, with Westerners believing in controllability more than Easterners.*
- *Tacit assumptions about stability vs. change, with Westerners seeing stability where Easterners see change.*
- *Preferred patterns of explanation for events, with Westerners focusing on objects and Easterners casting a broader net to include the environment.*
- *Habits of organizing the world, with Westerners preferring categories and Easterners being more likely to emphasize relationships.*
- *Use of formal logical rules, with Westerners being more inclined to use logical rules to understand events than Easterners*
- *Application of dialectical approaches, with Easterners being more inclined to seek the Middle Way when confronted with apparent contradiction and Westerners being more inclined to insist on the correctness of one belief vs. another.* (Nisbett, 2003)

If these observations have any basis in fact, we see that our concepts about the world leak into our management styles. We seek to control our environments, we prefer stability and categorization over change and relationship, and we are inclined to use logical rules in lieu of understanding events. Are these habits we need to help our people unlearn?

I personally believe it is possible to create an effective, Western-styled, KATA – even one specific to our own individual organizations. We may need to adjustment the way we approach creating a Lean culture, but one thing is certain. The onus of developing our IT organizational routines of thinking and practice, along with creating and maintain the organizational culture through consistent role modeling, teaching, and coaching is a leadership responsibility. And it is a must.

So now you have a book, patterns, examples and the know-how. Will your Lean IT implementation succeed?

... *I.T.* depends.

About the Author

Ray Jarrett is currently the Principle Strategist at Lean Strategies, LLC. Ray has held numerous management / executive management, IT, Supply Chain and Lean positions over the past 15 years of his career. Ray is a forward-thinker and possesses extensive knowledge and formal education in the areas of IT, Lean  Manufacturing, Six Sigma, Agile, CMMI, and TOGAF to name a few. Ray is an experienced lecturer and has an extensive track record of success and execution in both the corporate and entrepreneurial marketplaces. Ray is a U.S. Air Force veteran and attended college at the University of Central Oklahoma in Edmond, OK. Ray loves sports (golf!), is an accomplished musician, a voracious reader and enjoys an occasional game of chess and/or a Star Trek Episode.

"If you think in terms of a year, plant a seed; if in terms of ten years, plant trees; if in terms of 100 years, teach the people." **Confucius**

Recommendations

Ryan Follmer

Sr. Supervisor IT Support Services

"It's been a real pleasure working together with Ray Jarrett at ATK. I found him to be a person with strong IT background and solid recognition of business solutions. Ray is an insightful, energetic, passionate co-worker with the skill set to analyze and manage projects and uphold business objectives. Ray has been a pleasure to be associated with and always presents a positive approach to all his challenges. He has also served as a great mentor to me as I grew within the company. It is my pleasure to recommend him personally and professionally."

Ronald Gilmore
North America Sales Leader Software Industry Products, IBM
"Ray is a true professional who understands the value of a partnership. We worked with Ray to identify ways to increase productivity and reduce costs at ATK. Ray was always fair and honest in our dealings and looked for ways to add value to his firm while treating me and my team partners not vendors in working through the process. He is a change agent that can help transform an organization and will always add value while working through the process."

John Rassieur
Associate Partner, IBM - Global Business Services
"Ray is a very talented Supply Chain IT professional. He blends a broad understanding of the intricacies of supply chain planning and operations with enabling technology to deliver enterprise capable solutions. His ability to recognize the linkage between aspects of the supply chain and thus potential gaps is remarkable resulting in superior business solutions. Ray understands what it means to be a partner with service providers. He balances the relationship between his company and service providers with care and diplomacy."

Diane Condeni
Business Analyst, ATK Small Caliber Systems
"I reported to Ray during his tenure as IT Manager at ATK. Ray championed the importance of IT's alignment with the business, and the power of IT to help the business achieve their business-critical strategies. Ray is a forward-thinker. He was a proponent of, and first initiator of the Business Analysis model within ATK, a Fortune 500 company. Ray also developed and led the first cross-functional business unit team to integrate and prioritize large IT projects and initiatives at the Small Caliber Systems location. Ray's energetic and positive leadership style, strong IT and

business background, and excellent communication skills, are all assets which contribute to his successes."

Rudy Boone
Rudy hired you as a IT Consultant in 2005
Top qualities: **Great Results, Expert, On Time**

"Ray successfully led a project to help implement a wage transfer system to support the movement of 2,500 hourly employees between different jobs to support a $1 Billion operation. Ray's knowledge of effective project planning and management led to the completion of the project on time and within budget. Ray is a true business partner who can achieve results despite adversity."

Acknowledgements

I put off writing this part of the book at least two dozen times...

The most extensive amount of research and effort went into writing this part of the book although I know there is a chance no one will ever read it. I never gave the acknowledgements in a book much credence until now.

When I take inventory of my personal earthly possessions, I would have to say my most prized set of inanimate objects is my library. I have a collection of thousands upon thousands of books – literally every room in my house has a bookshelf or a corner crammed with books – and those are just the ones that are not in storage. My library covers every subject from quantum mechanics to religion; from Star Trek metaphysics to the classic Five Dialogues by Plato. I inherited both the books in my father's library and his love of the written *logos*.

Why is this important? The research. I read hundreds and hundreds of acknowledgements, and I have to tell you, not too many of them made a big impression on me. Some were in the front of the book, some in the back. Some were several pages long and some were less than five words long. Some books omitted them completely. Some were lacked imagination and most were obligatory. Half of the books were filled with list of names qualifying as a veritable "who's who" in the author's field of study. The others were filled with the names of enough family members to resemble a program at a reunion. Some were syrupy sweet and others were short and sweet. I even looked at movie and music credits. David Sanborn has dedicated every album he has ever recorded to his son, Jonathan. The "Bad Robot" running around in the meadow at the end of every J.J. Abrams movie is a tribute to his kids, Henry and Gracie.

I thought to go for the shock factor, but I've had enough of that for one lifetime! But…there is the Star Trek - TNG episode, *"Tapestry"* reminding me not to live my life with regrets. What to do?

Then revelation and inspiration struck. It occurred to me that for all of their hard work, towering intellect and lifelong endeavors, the thing that mattered most to the majority of authors was a handful of people whom the rest of the world would never know. And that seems cool – very cool.

I'm not sure how my acknowledgements will read in subsequent books, but I finally have this one figured out. I do know this – this is my own little space of the world. I will have to assume that almost no one will read it. But undoubtedly, one day, someone may see this – probably the person who has "turned on" the one DNA base pair out of the $\pm$ 3.2 billion pairs in the haploid human genome – the one that makes you read hundreds of acknowledgements in every book you can get your hands on. Sigh.

So in this book, I acknowledge two of the many people who have supported me unconditionally. Both would tell me that they are very proud of the person I have become and not just because I wrote a book. Nonetheless, I am eternally grateful for the wisdom and guidance from both of them over the years. They both provided me with the inspiration to write this book. Both live in my heart; one I have seen and miss and the other I have never seen. A very special thank you to both of my Fathers - my heavenly Father and my dad, Raymond J. Jarrett, Sr. (1939 – 2007).

Bibliography

Brittanica.com. (2009, May 9). Retrieved Mar 1, 2012, from Britannica Online Encyclopedia: http://www.brittanica.com

Chernev, I. (1992). *The Most Instructive Games of Chess Ever Played: 62 Masterpieces of Chess Strategy.* Mineola, New York: Dover Publications, Inc.

DeMenthe, B. L. (2011, October 15). *Kata.* Retrieved April 5, 2012, from Wikipedia: http://en.wikipedia.org/wiki/Kata

High, P. A. (2009). *World Class IT: Why Business Succeds When IT Triumphs.* San Francisco, CA: Jossey-Bass.

Ichijo, K. a. (2011, October 15). *Kata.* Retrieved April 5, 2012, from Wikipedia: http://en.wikipedia.org/wiki/Kata

Liker, J. (2004). *The Toyota Way: 14 Management Principles from the World's Greatest Manufacturer.* New York City: McGraw-Hill.

Lobi, V. (Director). (1998). *Star Trek Voyager, "The Omega Directive"* [Motion Picture].

Nisbett, R. (2003). *The Geography of Thought: How Asians and Westerners Think Differently...and Why*. New York: The Free Press.

Ohno, T. (1988). *Toyota Production System: Beyond Large-Scale Production.* Portland, Oregon: Productivity, Inc.

Reis, E. (2011). *The Lean Startup.* New York: Crowne Business.

Rother, M. (2010). *Toyota KATA - Managing People for Improvement, Adaptiveness, and Superior Results.* New York, Chicago, San Francisco, Lisben, London, Madrid, Mexico City, Milan, New Delhi, San Juan, Seoul, Singapore, Sidney, Toronto: McGraw Hill.

Schein, E. (2011, October 15). *Kata.* Retrieved April 5, 2012, from Wikipedia: http://en.wikipedia.org/wiki/Kata

Tangredi, D. (2005, January 29). *Insanity in the Sign & Graphics Industry*. Retrieved March 23, 2012, from Ezine Articles: http://EzineArticles.com/12047

Weill, J. W. (2011). Six IT Decisions Your IT People Shouldn't Make. In H. B. Press, *Aligning Technology with Strategy* (pp. 12,15). Boston, Massachutsetts 02163: Harvard Business School Publishing Corporation.

Wikipedia. (2011, October 15). *Kata*. Retrieved April 5, 2012, from Wikipedia: http://en.wikipedia.org/wiki/Kata

Wikipedia. (2012, January 21). *Lean IT*. Retrieved February 21, 2012, from Wikipedia The Free Encyclopedia: http://en.wikipedia.org/wiki/Lean_IT

Youngman, J. K. (2009). *A Guide to Implementing the Theory of Constraints (TOC)*. Retrieved April 3, 2012, from Theory of Constraints Production Batch Issues: http://www.dbrmfg.co.nz/Production%20Batch%20Issues.htm

Index

5

5S	ii, 22, 24
5-Whys	75

A

A3	45, 75, 88, 111
Access DB	44
Agile	50, 51, 59, 88
Albert Einstein	99
Anderson	**20, 50**
Andon	56
ANDON	43, 56, 61, 87, 92
automating chaos	*101*

B

Batch	**57**
burn down chart	42
Business Intelligence (BI)	105
Business Process Improvement	22

C

Capablanca	**101, 102**
Chakotay	**29**
Confucius	**12**
COTS	13, 78

D

DMS-250/300	104

E

ERP system	91, 105

Excel	44, 49

F

Finance	10, 103
FMEA	56
FPY	38, 39

G

Gemba Board	22, 63
Genba	43, 56, 58, 63
Geography of Thought	*111*

H

Harvard Business Review	*104*
Heijunka Board	47, 74, 75, 76
High	**7, 51**

I

Infrastructure Pattern Matching	32

J

Jeanne W. Ross	**104**
JIT43, 63, 75	

K

kaizen	ii, 22, 24, 75, 111
Kaizen	43, 63, 64, 66, 98
Kanban	43, 50
KATA	**107, 109, 112**
KTLO	11, 13, 14

L

LAN/WAN	66
Lean Culture	20, 21, 29, 59, 69, 78, 107
Lean IT	3
Lean IT blueprint	36
Liker	**20, 27**

M

Marshall	**101, 102**
Material Price Index	90
META Group	32
muda	29
Muda	43, 55, 63

N

Nemawashi	55, 61

O

Ohno	**22, 26**
On-Time Delivery	74
OTD	74, 75

P

P.I.C. Pattern	*33, 35, 54, 58, 77*
Pattern Ranking	34, 41, 54, 62
Perkerwicz	**107**
Peter Weill	**104**
poka-yoke	**50**
Poka-yoke	43
portal technology	*105*

R

R.P.M Pattern	*35*
Reinertsen	**20, 50**
Reis	**50**
Ries	**20**
Rother	**20, 110**

S

service integration hub	*105*
Service Level Agreements	59, 80
Seven of Nine	**29**
shop-floor data network	*105*
Six Sigma	22
Sprint	*104*
Steering Committee	*106*
Supply Chain	i, 10, 63, 90, 91, 92, 93, 103, 105
Sutton	**50**

T

T.O.N.E. Pattern	*33, 35, 62, 65, 90*
Takt time	42, 88
Takt Time	24, 42, 43
The Lean Startup	**20, 99**
The Toyota Way	**20, 27, 49**
TOGAF	11
Toyota	**20, 26, 27, 49, 101, 110**
Toyota KATA	**20, 110**
TRIZ	**56**

V

value stream mapping	39, 53

W

WCIT	8, 19, 36, 45, 46, 70
Wikipedia	4, 26, 108
World Class IT	**1, i, 7, 36, 51, 80**

Y

Youngman	57

Z

zugszwang	*101*

Need More Help?

For upcoming seminars, resources, consulting and more information about Lean Strategies, visit one of our social media sites below.

Social Media	QRC Code
Website – http://www.newleanstrategy.com	
Twitter – https://twitter.com/#!/rjjjazz	
Facebook - http://www.facebook.com/LeanStrategies	

www.ingramcontent.com/pod-product-compliance
Lightning Source LLC
Chambersburg PA
CBHW030810180526
45163CB00003B/1216